CAMBRIDGE LIBRARY COLLECTION

Books of enduring scholarly value

Earth Sciences

In the nineteenth century, geology emerged as a distinct academic discipline. It pointed the way towards the theory of evolution, as scientists including Gideon Mantell, Adam Sedgwick, Charles Lyell and Roderick Murchison began to use the evidence of minerals, rock formations and fossils to demonstrate that the earth was older by millions of years than the conventional, Bible-based wisdom had supposed. They argued convincingly that the climate, flora and fauna of the distant past could be deduced from geological evidence. Volcanic activity, the formation of mountains, and the action of glaciers and rivers, tides and ocean currents also became better understood. This series includes landmark publications by pioneers of the modern earth sciences, who advanced the scientific understanding of our planet and the processes by which it is constantly re-shaped.

The Complete Weather Guide

Early nineteenth-century farmers often sowed their crops on an arbitrarily chosen day every year. Impatient with this practice, naturalist Joseph Taylor (*c*.1761–1844) presents an alternative method in this work, which first appeared in 1812. He argues that by studying the atmosphere, the behaviour of animals and the condition of local flora, a farmer can not only determine the optimal time for sowing, but also forecast the weather. Including the Shepherd of Banbury's famous rules for judging changes in the weather, alongside remarks on the quality of this wisdom, Taylor's book also draws on a wealth of wider countryside knowledge. He observes, for example, that the flowering of primroses and lettuce occurs at such precise times as to be useful for botanical clocks, while the proximity of bees to their hives and the agitation of dogs suggest oncoming weather conditions.

Cambridge University Press has long been a pioneer in the reissuing of out-of-print titles from its own backlist, producing digital reprints of books that are still sought after by scholars and students but could not be reprinted economically using traditional technology. The Cambridge Library Collection extends this activity to a wider range of books which are still of importance to researchers and professionals, either for the source material they contain, or as landmarks in the history of their academic discipline.

Drawing from the world-renowned collections in the Cambridge University Library and other partner libraries, and guided by the advice of experts in each subject area, Cambridge University Press is using state-of-the-art scanning machines in its own Printing House to capture the content of each book selected for inclusion. The files are processed to give a consistently clear, crisp image, and the books finished to the high quality standard for which the Press is recognised around the world. The latest print-on-demand technology ensures that the books will remain available indefinitely, and that orders for single or multiple copies can quickly be supplied.

The Cambridge Library Collection brings back to life books of enduring scholarly value (including out-of-copyright works originally issued by other publishers) across a wide range of disciplines in the humanities and social sciences and in science and technology.

The Complete Weather Guide

A Collection of Practical Observations for Prognosticating the Weather, Drawn from Plants, Animals, Inanimate Bodies, and Also by Means of Philosophical Instruments

Joseph Taylor

CAMBRIDGE
UNIVERSITY PRESS

University Printing House, Cambridge, CB2 8BS, United Kingdom

Published in the United States of America by Cambridge University Press, New York

Cambridge University Press is part of the University of Cambridge.
It furthers the University's mission by disseminating knowledge in the pursuit of education, learning and research at the highest international levels of excellence.

www.cambridge.org
Information on this title: www.cambridge.org/9781108065313

This edition first published 1812
This digitally printed version 2013

ISBN 978-1-108-06531-3 Paperback

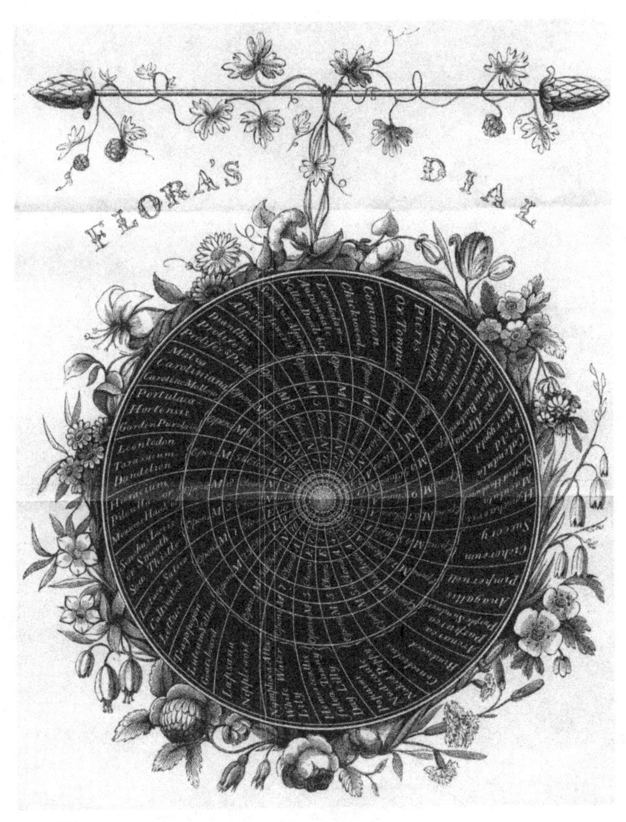

The material originally positioned here is too large for reproduction in this reissue. A PDF can be downloaded from the web address given on page iv of this book, by clicking on 'Resources Available .

THE

COMPLETE WEATHER GUIDE:

A COLLECTION OF

PRACTICAL OBSERVATIONS

FOR PROGNOSTICATING THE WEATHER,

DRAWN FROM PLANTS, ANIMALS, INANIMATE BODIES, AND ALSO BY MEANS OF PHILOSOPHICAL INSTRUMENTS;

Including

THE SHEPHERD OF BANBURY'S

RULES,

EXPLAINED ON PHILOSOPHICAL PRINCIPLES.

WITH

AN APPENDIX

OF MISCELLANEOUS OBSERVATIONS ON METEOROLOGY,

A CURIOUS

BOTANICAL CLOCK,

&c. &c. &c.

By JOSEPH TAYLOR.

LONDON:

PRINTED FOR JOHN HARDING, 36, ST. JAMES'S STREET.

1812.

Harding and Wright, Printers, St. John's Square, London.

TABLE

OF

CONTENTS.

PART III.

THE SHEPHERD OF BANBURY'S RULES, BY WHICH TO JUDGE OF THE CHANGES OF THE WEATHER, (GROUNDED ON FORTY YEARS' EXPERIENCE,) METHODIZED AND ARRANGED UNDER DISTINCT HEADS.

APPENDIX.

PREFACE.

THE phenomena of the weather have, at all times, attracted much of the attention of mankind; because their subsistence and their comfort in a great measure depended on them. It was not, however, until the seventeenth century, that any considerable progress was made in investigating the laws of meteorology;—a subject, of all others, the most interesting to farmers, and to the agricultural interest in general. Since that time, philosophers have been enabled to make numerous and accurate meteorological observations; which have been collected from time to time, and many important practical results have been deduced therefrom.

The design of the present publication is, to collect such observations as may be depended on, and to bring together such a variety of important information on the state of the weather, as may enable those, who are interested in agricultural pursuits, profitably to regulate the management and housing of their crops.

The Work consists of three principal parts.

I. Observations indicating the probable changes of the weather from the appearances of nature.

II. Observations prognosticating such changes, from philosophical instruments.

III. "The Shepherd of Banbury's rules to judge of the changes of the weather."

Who the shepherd of Banbury was, we know not; nor indeed have we any proof that the rules called his, were penned by a real shepherd: both these points are however immaterial: *their truth is their best voucher.* Mr. Claridge, (who published them in the year 1744,) states, that they are grounded on forty years experience, and thus, very rightly, accounts for the

presumption in their favour. " The shep-herd," he remarks, " whose sole business it is
" to observe what has a reference to the flock
" under his care, who spends all his days, and
" many of his nights in the open air, under the
" wide-spread canopy of Heaven, is obliged to
" take particular notice of the alterations of the
" weather; and when he comes to take a pleasure
" in making such observations, it is amazing how
" great a progress he makes in them, and to how
" great a certainty he arrives at last, by mere
" dint of comparing signs and events, and cor-
" recting one remark by another. Every thing,
" in time, becomes to him a sort of weather-gage.
" The sun, the moon, the stars, the clouds, the
" winds, the mists, the trees, the flowers, the
" herbs, and almost every animal with which he
" is acquainted, all these become, to such a per-
" son, instruments of real knowledge."

As the shepherd's rules were originally published without much regard to order, they are now methodized under distinct heads, and are either confirmed by facts and collateral observations, or are explained on the principles of the latest discoveries; with which the original editor appears to have been unacquainted, or omitted to notice.

An Appendix is subjoined, containing some miscellaneous hints, which are not strictly referable to either of the preceding heads: a copious Index closes the work, which is now, with deference, offered to the attention of a liberal public in general, and more especially to the agricultural interest.

INTRODUCTION.

THE advantages arising from a foreknowledge of the changes of the weather, were duly appreciated by the ancients, in whose writings many valuable hints have been preserved: and, though ill-founded predictions, in more recent times, have cast some discredit on the study of the changes of the weather; yet it is evident to the diligent observer of nature, that a considerable degree of certainty is attainable, both in tracing the causes, and also in foreseeing the succession of those changes. The experienced fisherman, from his constant observance of the sky, will rarely unfurl his sails when a storm is approaching; and, in like manner, if farmers were equally attentive, and had acquired equal judgment in this art, they would as seldom be overtaken by unexpected changes.

They must not however at all times look so high, as to neglect what passes around them on the surface of the earth. The vegetation of plants, especially of the natives of each country, is a kalendar well worthy of observation, as a directory of the seasons proper for certain works in the spring: nor should the accidents which happen to even the least useful plants be neglected, because they may afford hints of what should be done to prevent similar evils in those of greater utility.

Linnæus and his disciples have given excellent instructions on this head. One of them in particular, Mr. Harold Barck*, states that it was then the fourth year since that illustrious botanist exhorted his countrymen to observe with all care

* Dissertation on the Foliation of Trees, presented in 1753 to the university of Upsal, then under the presidency of Linnæus himself.

and diligence, at what time each tree expands its buds, and unfolds its leaves; justly conceiving that his country, (and the remark is applicable to every other,) might reap some benefit, from similar observations made in different places. As one of the apparent advantages, he advises the prudent husbandman to watch with the greatest care the proper time for sowing; because this, with the divine assistance, produces plenty of provision, and lays the foundation of the public welfare of the state, and of the private happiness of the people. The ignorant farmer, continues he, tenacious of the ways and customs of his ancestors, fixes his sowing-season generally to a month, and sometimes to a particular day, without considering whether the earth be duly prepared to receive the seed: hence it frequently happens, that the fields do not yield a produce correspondent to his sanguine expectations. The wise economist should therefore fix certain signs by which to judge of the proper time for sowing. We look up to the stars, and, without reason, suppose that the changes on earth will answer to the heavenly bodies; entirely neglecting the things which grow around us. We see trees open their buds, and their leaves expand, whence we conclude that the spring is approaching, and experience supports us in the conclusion; but no one has yet been able to shew *what* trees Providence intended should be our kalendar, so that we might ascertain on what day the countryman ought to sow his grain. Neither can it be denied but that the same power, which brings forth the leaves of trees, will also cause the grain to vegetate: nor can any one justly assert that a premature sowing will uniformly accelerate a ripe harvest. No means therefore seem to promise success so much, as the taking of our rule for sowing from the leafing of trees. With this view it must be observed in what order every tree puts forth its leaves, according to its species, the heat of the atmosphere, and the quality of the soil. Afterwards by comparing together the observations of several years, it will not be difficult to determine, from the foliation of trees, (if not certainly, at least probably); the time when annual plants ought to be sown. It will be necessary likewise to remark what sowings made in different parts of the spring produce the best crops, in order that by comparing these with the leafing of trees, it may appear which is the most proper time for sowing: nor will it be amiss in like manner to note at what times certain plants, especially the most remarkable in every province or country, blow; in order that it may be known whether the year makes a quicker or slower progress.

Linnæus's method of carefully observing the foliation of trees, &c. would undoubtedly determine the proper time for spring-sowing; and Pliny, after mentioning the several constellations by which farmers were guided in his time, instructs the husbandman with regard to autumnal sowing, upon a principle similar to that of our great modern naturalist. "Why," says he,*
"does the husbandman look up to the stars, of
"which he is ignorant, whilst every hedge and
"tree point out the season by the fall of their
"leaves? This circumstance will indicate the
"temperature of the air in every climate, and
"shew whether the season be early or late.
"This constitutes an universal rule for the

* Nat. Hist. b. xviii. ch. 25.

" whole world; because trees shed their leaves
" in every country according to the difference
" of the seasons. This gives a general signal
" for sowing; nature declaring, that she has
" then covered the earth against the inclemency
" of the winter, and enriched it with this ma-
" nure."

An accurate observer of nature, (the late Mr. Stillingfleet,) has related, that he himself was told by a common husbandman in Norfolk, that when the oak catkins begin to shed their seed, it is a proper time to sow barley: " And why," adds he, very properly, " may not some other
" trees serve to direct the farmer for the sowing
" of other seeds? The prudent gardener never
" ventures to put his house plants out till the mul-
" berry leaf is of a certain growth." Hesiod, he continues,* began to fix the proper season for ploughing, sowing, &c. by the appearance of birds of passage, or of insects, or by the flowering of plants; but we have no record of observations of this kind being made till Linnæus wrote. Hesiod says, that when the voice of the crane is

* Miscell. Tracts, 8vo. p. 147.

heard over-head, then is the time for ploughing; that if it should happen to rain three days together when the cuckow sings, late sowing will then be as good as early sowing; that when snails begin to creep out of their holes, and climb up plants, it is time to cease digging about the vine.

There is a wonderful coincidence, which probably takes place in all countries, between vegetation and the arrival of certain birds of passage. Linnæus says, that the *wood-anemone* (in Sweden) blows from the time of the arrival of the swallow; and Mr. Stillingfleet finds by a diary which he kept in Norfolk for the year 1755, that the swallow appeared there on the 6th of April, and the *wood-anemone* was in bloom on the 10th of the same month. Linnæus observes, that the *Marsh-marigold* blows when the cuckow sings; and Mr. Stillingfleet finds by his diary, that the *Marsh-marigold* was in blossom on the 7th of April, and the cuckow sung the same day.

The methods, here hinted at, deserve the most serious attention: a series of similar observations, properly made by intelligent persons, in different parts, and afterwards rightly compared and combined, would soon afford almost infallible rules to guide the husbandman in one of the most important parts of agriculture.

The principal points necessary in making such observations are, 1st, That they be continued for a due length of time, and that the *time* and *place* of observation be particularly specified: 2dly, That they may be made on the same *subjects:* and 3dly, That the *soil* and *exposure* be carefully noticed and described, in order to their being duly compared with the field intended to be sown. The necessity of being as exact as possible in this last article, will appear to every one who does but consider, what all know, that the *north-wind*, *shade*, and a *moist soil*, hinder the leafing of trees, as much as a *dry situation* on the *slope* of a hill inclining to the *south* promotes it.—Another circumstance which would greatly facilitate the application of these observations, is, to take the trees in their progressive order of leafing: for nature is always regular, and the guide would then be sure.

Among the various phenomena, which attentive observers have found to indicate the approaching changes in the atmosphere, the following may be considered as affording the most certain signs.

I. From Vegetables.

II. From Animals.

III. From the Atmosphere.

IV. From the Seasons.

V. From appearances presented by philosophical instruments, which have been invented for the express purpose of exhibiting the state of the weather, and its approaching variations.

Each of these articles, it is attempted to illustrate in the following sections; which will be terminated by some miscellaneous information, not strictly referable to either of the former heads.

A

WEATHER GUIDE,

&c.

PART I.

OBSERVATIONS, BY WHICH TO JUDGE OF THE CHANGES OF THE WEATHER, DEDUCED FROM THE APPEARANCES OF NATURE.

SECTION I.

CHANGES OF WEATHER, INDICATED BY VEGETABLES.

Chickweed—is an excellent out-of-door barometer. When the flower expands boldly and fully, no rain will happen for four hours or upwards: if it continues in that open state, no rain will disturb the summer's day: when it half conceals its miniature flower, the day is generally showery; but if it entirely shuts up, or veils the white flower with its green mantle, let the traveller put on his great coat, and the ploughman with his beasts of draught, expect rest from their labour.

SIBERIAN SOW THISTLE.—If the flowers of this plant keep open all night, rain will certainly fall the next day.

The different species of TREFOIL always contract their leaves at the approach of a storm:—so certainly does this take place, that these plants have been termed "*The Husbandman's Barometer.*"

AFRICAN MARYGOLD.—If this plant opens not its flowers in the morning about seven o'clock, you may be sure it will rain that day, unless it thunders.

The CONVOLVULUS also, and the PIMPERNEL (*Anagallis*), fold up their leaves on the approach of rain, the last in particular is termed the poor man's weather glass.

WHITE THORNS and DOG-ROSE BUSHES.—Wet summers are generally attended with an unusual quantity of seed on these shrubs: whence their unusual fruitfulness is a sign of a severe winter.

Beside these, there are several plants,* especially those with compound yellow flowers, which nod and during the whole day turn their flowers towards the sun; viz. to the east in the morning, to the south at noon, and to the west towards evening; this is very observable in the *sonchus arvensis* or SOW THISTLE:—and it is a well known fact, that a great part of the plants in a serene sky expand their flowers, and as it were with cheerful looks behold the light of the sun; but before rain they shut them up, as the tulip.—The flowers of the *draba alpina*, alpine whitlow grass, the *parthenium, foliis ovatis crenatis*, or bastard feverfew with egg-shaped crenated leaves, and the *trientalis* or winter-green, hang down in the night, as if the plants were asleep, lest rain or the moist air should injure the fertilizing dust.—The trefoils, and one species of wood sorrel, also shut up or double their leaves before storms and tempests, but

* We are indebted for these remarks to Dr. Thornton's splendid "Illustration of the Sexual System of Linnæus."

in a serene sky expand or unfold them, so that the husbandman can pretty clearly foretel tempests from them.—It is also well known that the *bauhinia*, or mountain ebony, sensitive plants and *cassia*, observe the same rule.

SECTION II.

CHANGES OF WEATHER, INDICATED BY ANIMALS.

Preliminary Observations.

THE fluids and solids of organised beings, and their animal machines, are constructed in such a manner, that a certain degree of motion puts them in a good state, while an augmentation or diminution of it deranges or destroys that state. The fluids (which by their nature are easily moved,) as well as the fibres (which are highly susceptible of irritation,) are readily affected by changes of the surrounding atmosphere, and suffer from the impression, whether the air varies in its weight, or qualities, or is changed in regard to its elasticity. Among those who are sound and in perfect health, we find vivacity, good spirits, and great agility, when the air is pure and elastic; on the other hand, when the air becomes light and damp, and is deprived of its elasticity, it throws the body into a state of languor and debility. Valetudinarians, whose constitutions are delicate, or who are advanced in life, are much sooner sensible of the impressions occasioned by the weather, than those who are strong and robust. In general the senses of men, who in their way of life deviate from the simplicity of nature, are coarse, dull, and void of energy. Those also, who are distracted by a thousand other objects, scarcely feel the impression of the air, and if they speak of it to fill up a vacuum in their miserable and frivolous conversation, they do it without thinking of its causes or effects, and without ever paying attention to them. But animals,—which retain their natural instinct, which have their organs better constituted, and their senses in a more perfect state, and besides are not changed by vicious and depraved habits,—perceive sooner, and are more susceptible of the impressions produced in them by variations of the atmosphere, and sooner exhibit signs of them.

Until the discovery of animal electricity, little attention was paid to those signs, which were consequently ascribed to a certain natural prescience. But, as the electric matter issuing from the earth diffuses itself through the atmosphere, it must penetrate and agitate the frail machines in question, and as it carries with it vapours and exhalations of various kinds, these must produce, on machines so delicate, different sensations, which make them move in a different manner; and, according as they receive impressions agreeable or troublesome, they exhibit signs of joy or sadness; send forth cries, or are silent; move or remain at rest, as is observed in all kinds of animals, without excepting man, when the weather is about to change.

In the last place, internal and animal electricity, which in all probability is the agent of life, and the grand source of organic motion, must be as much subject to modifications as the external electricity, from which it acquires new force and activity, by the vapours and humidity of the atmosphere, which absorbing the electric matter in abundance, or serving it as a conductor, draw it off from the animal machine. Hence arises that languor and debility, which are experienced during wet weather and when the south winds prevail; and for the same reason, the moisture which has penetrated the organs, at least such as are weak, or have suffered any hurt or injury, or been exposed to some new agitation, produces uneasiness, and occasions pain. It is difficult to explain clearly and with precision how all this takes place; that is, how the electricity is excited, and by what mechanism exhalations and vapours affect animals, and produce changes in their bodies, since we are not acquainted with the curious organisation of the most delicate parts of these machines; but we can observe and perceive the progress of these phenomena, as well as those by which they are produced.

Common and familiar Signs exhibited by Animals, which indicate approaching Changes of the Weather.

1. When bats remain longer than usual abroad from their holes, fly about in greater numbers, and to a greater distance than common, it announces that the following day will be warm and serene; but if they enter their houses, and send forth loud and repeated cries, it indicates bad weather.

2. If the owl is heard to scream during bad weather, it announces that it will soon become fine.

3. The croaking of crows indicates fine weather.

4. When the raven croaks three or four times, extending his wings and shaking the leaves, it is a sign of serene weather.

5. It is an indication of rain and stormy weather, when ducks and geese fly backward and forward; when they plunge frequently into the water, or send forth cries, and fly about.

6. If bees do not remove to a great distance

from their hives, it announces rain; if they return to their hives before the usual time, it may be concluded it will soon fall. On the contrary, if they fly far from their hives, and return home late, they foretel very *fair* and *hot* weather.

7. If pigeons return slowly to the pigeon-house, it indicates that the succeeding day will be rainy.

8. It is a sign of rain or wind when sparrows chirp a great deal, and make a noise to each other to assemble.

9. When fowls and chickens roll in the sand more than usual, it announces rain:—so, if poultry go to roost;—if tame fowls grub in the dust and clap their wings, small birds seem to duck and wash in the sand;—if cocks crow late and early, or at uncommon hours, clapping their wings;—if the red-breast be seen near houses; —all these are indications that rain is not far distant.

10. Peacocks, which cry during the night, have a presentiment of rain.

11. It is believed to be a sign of bad weather when the swallows fly in such a manner as to

brush the surface of the water, and to touch it frequently with their wings and breast.

12. The weather is about to become cloudy, and to change for the worse, when flies sting, and are more troublesome than usual.

13. When gnats collect themselves before the setting of the sun, and form a sort of vortex in the shape of a column, it announces fine weather. If they play up and down in the open air near sun-set, they presage *heat;* if in the shade, warm and mild showers; but if they join in stinging those who pass by them, *cold weather* and much rain may be expected.

14. When sea-fowl and other aquatic birds retire to the sea shore or marshes, it indicates a change of weather, and a sudden storm.

15. If cranes fly exceedingly high, in silence, and ranged in order, it is a sign of approaching fine weather; but, if they fly in disorder, or immediately return with cries, it announces wind. —The appearance also of cranes and of other birds of passage early in autumn, announces a severe winter; for it is a sign that it has already begun in the northern countries.

16. If larks rise very high, and continue to sing for a long time;—also, if kites fly aloft,—these are signs of *fair* weather.

17. When dolphins sport and make frequent leaps, the sea being tranquil and calm, it denotes that the wind will blow from that quarter from which they proceed.

18. If frogs croak more than usual;—if toads issue from their holes in the evening in great numbers;—if earth-worms come forth from the earth;—if ants remove their eggs from their small hills;—if moles throw up the earth more than usual;—if asses shake and agitate their ears, and bray more frequently than usual;—if hogs shake and spoil the stalks of corn;—if bats send forth cries, and fly into the house;—if dogs roll on the ground, and scratch up the earth with their fore-feet;—if cows or oxen look towards the heavens, and turn up their nostrils as if catching some smell;—if oxen lick their fore-feet, and if oxen and dogs lie on their right side;—if rats and mice are more restless than usual; all these are signs which announce *rain*.

19. The case is the same when animals crowd together.

20. As soon as bad weather approaches, the ass will hang down his ears forward, walk more slowly than usual, and rub himself against walls.

21. When goats and sheep are more obstinate and more desirous to crop their pastures, and seem to quit them with reluctance, and when the birds return slowly to their nests, rain may be soon expected.

22. An intelligent observer of Nature remarks——"That before a change of weather, on going to a sheep-fold, he has noticed these otherwise still and patient creatures, running about in different directions, jumping from the ground, and in their gambols apparently fighting; and, previous to a deep fall of snow, they will clear the ground of every scrap of turnip, or wisp of hay, within their reach; and retire, with accurate precision, for shelter, always to the spot which is best able to afford it."

23. A beautiful insect called the clock beetle, which flies about in the summer evenings in a

circular direction, with a loud buzzing noise, is said to foretel a fine day. It was consecrated by the Egyptians to the sun: the body is often coloured with a blueish or greenish gloss, sometimes brassy beneath.

24. The Leech.—Put a leech into a large phial three parts full of clear rain water, regularly change the same thrice a week, and let it stand on a window frame fronting the north. In fair and frosty weather it will be motionless, and rolled up in a spiral form, at the bottom of the glass; but prior to rain or snow, it will creep to the top, where, if the rain will be heavy, and of some continuance, it will remain a considerable time; if trifling, it will descend. Should the rain or snow be accompanied with wind, it wil dart about its habitation with an amazing celerity, and seldom ceases until it begins to blow hard. If a storm of thunder or lightning be approaching, it will be exceedingly agitated, and express its feelings in violent convulsive starts at the top of the glass. It is remarkable, that however fine and serene the weather may be, and not the least indication of a change, either

from the sky, the barometer, or any other cause whatever, yet if the animal ever shifts its position, or moves in a desultory manner, the coincident results will certainly occur within thirty-six hours; frequently within twenty-four, and sometimes in twelve; though its motions chiefly depend on the fall and duration of the wet, and the strength of the wind.

25. In men, frequently, aches, pains, wounds, and corns, are more troublesome, either towards rain or towards frost.

26. Persons of a plethoric (or full) habit of body are frequently oppressed with *drowsiness* and *heavy* sleep before rain falls.

SECTION III.

CHANGES OF WEATHER, INDICATED FROM THE APPEARANCES OF THE ATMOSPHERE, THE EARTH, SEASONS, &c.

EVAPORATION is the conversion of fluids, principally of water, into vapour; which, becoming specifically lighter than the atmosphere, is raised considerably above the surface of the earth, and afterwards by a partial condensation forms clouds. The heat of the sun, together with that of the electrical matter arising from the earth, is (according to many eminent philosophers) the cause of what is termed *Spontaneous Evaporation*, in order to distinguish it from that which is produced by artificial means. Evaporation is one of the great chemical processes, by which Nature supplies the whole vegetable kingdom with the dew and rain necessary for its support: hence it takes place at all times, not only from the surface of the ocean, but also from that of the earth. Nor is it confined to these: it is even carried on from the leaves of trees, grass, &c. with which the earth is covered. Great part of the water, which is thus raised, descends again during the night in the form of dew, being absorbed by those vegetables which yielded it before. One of the most beneficial effects of evaporation is, to cool the earth, and prevent it from being too much heated by the sun.

§ 1. *Changes indicated by the Clouds.*

1. THE clouds, called *Cirrus*, appear early after serene weather: they are, at first, indicated by a few threads pencilled as it were on the sky; these increase in length, and new ones are, in the mean time, added laterally. Often the first formed threads serve as stems to support numerous branches, which in their turn give rise to others. Their duration is uncertain, varying, from a few minutes after their first appearance, to an extent of many hours. It is long when they appear alone, and at great heights; and shorter when they are formed lower, and in the vicinity of other clouds. This modification, although in appearance almost motionless, is intimately connected with the variable motion of the atmosphere; and clouds of this kind have long been deemed a prognostic of the wind.

2. In fair weather, with light variable breezes, the sky is seldom quite clear from small groups of the oblique cirrus, which frequently come on from the leeward, and the direction of their in-

crease is *to* the windward. Continued wet weather is attended with horizontal sheets of this cloud, which subside quickly, and pass to the cirro-stratus. The cirrus pointing upward, is a distant indication of rain; and downward, a more immediate one of fair weather. Before storms they appear lower and denser, and usually in the quarter opposite to that from which the storm arises. Steady high winds are also preceded and attended by streaks running quite across the sky, in the direction they blow in. These, by an optical deception, appear to meet in the horizon.

3. The shooting or falling star, precedes a change of wind.

4. If clouds appear gradually to diminish, and dissolve into the air, so as to become invisible, it is an indication of fine weather.

5. If the sky, after being for a long time serene and blue, become *fretted* and spotted with small undulated clouds, not unlike the waves of the sea, rain will speedily follow.

6. It not unfrequently happens that two different currents of clouds appear: these are

certain signs of rain, particularly if the lower current fly swiftly before the wind. Should two such currents appear during summer, or hot weather, they announce a speedy thunder-storm.

7. Previously to heavy rains, especially at the approach of a thunder-storm, each cloud becomes larger than the former; and all are visibly increased in size.

8. When the solar rays break through the clouds, and are visible in the air, it shews that the atmosphere is filled with vapours, which will speedily be converted into rain.

9. If clouds are formed like fleeces, deep and dense (or thick and close) towards the middle, the edges being very white, while the surrounding sky is very bright and blue, they are of a frosty coldness, and will speedily fall (according to the season) either in hail, snow, or hasty showers of rain.

10. So, if clouds appear high in the air, in thin white trains, like locks of wool, or horses' tails, they indicate that the vapours are spread and scattered by contrary winds above; and

that a storm of wind, probably accompanied by rain, will soon blow below.

11. When the air is hazy, so that the solar light fades gradually, and looks white, rain will most certainly follow. In like manner, if the moon and stars grow dim in the night, and the air also be hazy, and a *halo*, *ring*, or *burr*, appear round the moon, it is a sure sign of rain.

12. If, in a very wet season, the sky is tinged with a sea-green colour, near the bottom, where it ought to be blue, it shews that rain will speedily follow and increase: when it is of a deep dead blue, it is overcharged with vapours, and the weather will be showery.

13. When the sun appears white at the setting, or goes down into a bank of clouds, which lie in the horizon, they indicate the approach or continuance of bad weather.

14. When it rains with an east wind, it will probably continue for twenty-four hours.

15. The heaviest rains, when of long continuance, generally begin with the wind blowing easterly, which gradually veers round to the

south; and the rains do not cease, until the wind has got to the west, or a little north-west.

16. While rain is falling, if any small space of the sky be observable, it is almost a certain sign that the rain will speedily cease.

17. If the clouds, that move with the wind, become stationary when they arrive at that part of the horizon which is opposite to the wind, and appear to accumulate, they announce a speedy fall of rain.

18. A rain-bow in the morning
Is the shepherd's warning:
But a rain-bow at night
Is the shepherd's delight.

19. If a rain-bow appear in *fair* weather, foul will follow; if in foul, fair will follow. A *double rain-bow* indicates much rain.

20. Synopsis of the colours of the rain bow.—The *purple* shews wind and rain; the *dark-red*, tempestuous;—the *light-red*, wind;—the *yellow* shews dry weather;—the *green* denotes rain;—the *blue*, that the air is clearing. By a careful observance of these colours, it may easily be calculated what weather will follow.

21. If an *Aurora Borealis* appear after several warm days, it is generally succeeded by a coldness of the air.

22. If the *Aurora Borealis* has been considerable, either an increased degree of cold is immediately produced, or bodies of clouds are formed.

---

§ 2. *The Nature of the four principal Winds, and their Effects.*

1. *Subsolanus*, or the east wind, is hot and dry, temperate, sweet, pure, subtle, and healthful; especially in the morning when the sun rises, by whom it is made more pure and subtle, expelling all infection.—The hoar-frost, which is first occasioned by the east wind, indicates that the cold will continue a long time—as was the case in the year 1770.—This is the driest wind, because it comes across the vast continent of Asia, which is but little watered by rivers or seas.

2. *Zephyrus*, or the west wind, is temperate, hot, moist and wholesome, especially in the evening; it dissolves the frost, ice, and snow, and causes the flowers and grass to spring; according to some, it produces thunder. It often blows rain, as it crosses the great Atlantic ocean, and attracts a great quantity of vapours.

3. *Septentrio*, or the north wind, is, for the most part, cold and dry, repelling moisture and rain; and though it causes cold and numbness, so nipping the fruits of the earth, and many times the forward buds of the spring, yet it drives away infections and noisome airs, and this is conducive to the preservation of health. Coming from the frigid zone, this is the coldest of all the four winds.—When north-west and south-east winds prevail together at two different heights in the atmosphere, if the south-east be the lower one, we may expect that the weather will become clear: the contrary will take place, if the south-east wind be highest.

4. *Auster*, or *Notus*, the south wind, is hot and moist, breeding thick clouds and sickness. This is the warmest, as it comes from the torrid

zone: a south-west wind most frequently brings rain.

5. A frequent change of wind, accompanied with an agitation of the clouds, denotes a sudden storm.

6. A fresh breeze generally springs up before sun-set, particularly in the summer.

7. The weather usually clears up at noon; but, if it rain at midnight, it seldom clears up till sun-set.

8. The winds, which begin to blow in the day-time, are much stronger, and endure longer, than those which begin to blow only in the night. Violent winds usually abate towards sun-set.

9. If the wind veer about uncertainly to various points of the compass, it is a sure sign of rain.

10. A howling or whistling wind denotes rain.

11. If the wind follow the sun's course, fair weather will follow.

12. Weather—either good or bad, which takes place in the night-time, is not, in general, of long duration; and, for the most part, wind is more uncommon in the night than in the day

time. Fine weather in the night, with scattered clouds, does not last.

13. Violent winds prevail more in the vicinity of mountains than in open plains.

14. A Venetian proverb says, "That the sudden storm from the north does not last three days."

15. If it thunders in the month of December, moderate and fine weather may be expected.

16. If it thunders at intervals in the spring-time, before the trees have acquired leaves, cold weather is still to be expected.

17. Thunder in the morning denotes wind at noon; in the evening rain and tempest.—If in summer there be no thunder, it may be expected that the ensuing winter will be sickly.—If it *lightens* on a clear star-light night, in the south or south-east, rain and wind will follow; if it lighten in an evening towards the north, south, or south-west, it indicates wind.

18. Hot weather generally precedes thunder, which is followed by cold showery weather.

19. If the wind does not change, the weather will continue the same.

20. When the wind is south-west during summer or autumn, and the temperature of the air is unusually cold for the season, both to the feeling and the thermometer, with a low barometer, much rain is to be expected.

21. Violent temperatures, as storms or great rains, produce a sort of crisis in the atmosphere, which produces a constant temperature, good or bad, for some months.

§ 3. *Other signs announcing changes of the weather, from the appearance of the earth, &c.*

1. A want, or too great a quantity of dew, being a mark of strong evaporation, announces rain: the case is the same with thick white hoar frost, which is only dew congealed.

2. On the other hand, when the weather inclines to rain, the water is seen to diminish in vases and fountains: because the humidity is then carried away by the evaporation of the electric matter.

3. In a morning, if a mist, which hangs over the low lands, draws towards the high lands, it is a sign of an approaching fine day.

4. If in the evening a white mist spread over a meadow through which a river flows, it will be drawn up by the sun, on the following morning, and a fine clear day will follow.

5. When the dew lies plentifully upon the grass after a fine day, another fine day may be expected; but if, after such a fair day, no dew fall nor any breeze be stirring, it indicates that the vapours are ascending, and will soon be precipitated in the form of rain.

6. It is certainly a surprising phenomenon to see the earth, after very long and very abundant rains, to be sometimes almost dry; the roads quite free from dirt, and the lands to become arid and parched. This is a sign that the rain has not altogether ceased, and denotes a continual efflux of electric matter, which being renewed carries with it, in the form of vapours, all the moisture that falls on the earth.

7. There is sometimes, however, a great deal of dirt, even after a very moderate rain, which,

in that case, is a sign of fine weather, because it indicates that evaporation has ceased. Dry stones, and moist earth, announce fine weather; dry earth, and moist stones, announce rain.

8. If the flame of a lamp crackles or flares, it indicates rainy weather.

9. The case is the same when soot detaches itself from the chimney and falls down.

10. It is a sign of rain, also, when the soot collected around pots or kettles takes fire, in the form of small points like grains of millet, because this phenomenon denotes that the air is cold and moist.

11. If the coals seem hotter than usual, or if the flame is more agitated, though the weather be calm at the time, it indicates wind.

12. When the flame burns steady, and proceeds straight upwards, it is a sign of fine weather.

13. If the sound of bells is heard at a gr at distance, it is a sign of wind, or of a change of weather.

14. The hollow sounds of forests; the murmuring noise of the waves of the sea, their foaming, and green and black colour, announce a storm.

15. Good or bad smells, seeming as if they were condensed, are a sign of a change of weather; either because exhalations arise and are dispersed in more abundance, which is a sign of an increase of elasticity; or because the air does not dispel or raise these exhalations, which indicates that the constitution of the atmosphere is motionless, light and void of elasticity.
trees, are agitated without any sensible wind, it is a sign of wind, and perhaps of rain; because it denotes that strong and penetrating exhalations arise from the earth.

17. These signs are less equivocal when the dry leaves and chaff are raised into a vortex, and carried into the air.

18. If salt,* marble, and glass become moist some days before rain;—if articles of wood, doors, and chests of drawers swell;—if the corns on the feet, and scars of old wounds, become painful;—all these signs indicate that aqueous vapours are exhaled from the earth, and are, no

* This may be ascertained in the following manner:—take a good pair of scales, in one of which let there be a

doubt, directed by the electric matter which diffuses itself there in greater abundance, and penetrates every body. Hence it happens that stones become moist, that wood swells, and salt becomes deliquescent by the moisture. When the stones, after being moist, become dry, it is a sign of fine weather.

---

§ 4. *Indications, afforded by an attentive observance of the seasons.*

1. If the earth and air abound with insects, worms, frogs, locusts, &c. ;—if the walnut-tree has more leaves than fruit ;—if there are large quantities of beans, fruit and fish ;—if the spring and summer are too damp ;—if hoar frost, fogs and

brass weight of one pound, and in the other a pound of salt, or salt petre well dried ; and place a stand beneath the scale, so as to prevent it from falling too low. When rain is approaching, the salt will swell and sink the scale ; and, as the weather grows fair, the brass will regain its ascendancy.

dew come on, at times when they are not generally seen, the year will be barren; the opposite signs announce fertility and abundance. Animals seem also to foresee and prognosticate fertility and barrenness:—It is said, that when the birds flock together, quit the woods and islands, and retire to the fields, villages and towns, it is a sign that the year will be barren.

2. A great quantity of snow in winter promises a fertile year; but abundant rains give reason to apprehend that it will be barren. A winter, during which a great deal of snow and rain falls, announces a very warm summer. It is generally believed, that thunder and storms in winter prognosticate abundance, because they fertilize the earth. When the spring is rainy, it produces an abundant crop of hay, and useless herbs; but, at the same time, a scarcity and dearth of grain. If it is warm, there will be plenty of fruit; but they will be almost all spoilt. If it is cold and dry, there will be few fruit or grapes; and silk worms will not thrive. If it is only dry, there will be few fruits, but they will be good. In the last place, if it is cold, the fruit will be late in coming to maturity.

3. A cold and windy May is favourable to corn.

4. If the spring and summer are both damp, or even both dry, a scarcity and dearth of provisions is to be apprehended. If the summer is dry there will be little corn, diseases will also prevail; but they will be more numerous if it is warm. If it is moderately cold, the corn will be late; but there will be a great deal of it, and the season will occasion few diseases.

5. A fine autumn announces a winter during which winds will prevail; if it is damp and rainy, it spoils the grapes, injures the sown fields, and threatens a scarcity. If it be too cold, or too warm, it produces maladies. A long severity of the seasons, either by winds, drought, dampness, heat or cold, becomes exceeding destructive to plants and animals. In general, there is a compensation for rain or drought between one season and another. A damp spring or summer is commonly followed by a fine autumn. If the winter is rainy, the spring will be dry; and if the former is dry, the latter will be damp. When the autumn is fine, the spring will be rainy. A

moist autumn, with a mild winter, is generally followed by a cold and dry spring, which greatly retards vegetation. Such, according to M. du Hamel, was the year 1741.—A severe autumn announces a windy winter.

6. Intervals of clear and pleasant weather often occur in November; and in general, the autumnal months are softer and less variable, than the correspondent ones in spring.

The above alternations have, in general, been clearly proved by the verity of a journal carried on for forty years.

---

§ 5. *Observations on the influence of the moon on the weather.*

The influence of the moon on the weather has, in all ages, been believed by the generality of mankind: the same opinion was embraced by the ancient philosophers, and several eminent philosophers of later times have thought the opinion not unworthy of notice. Now, although the moon only acts (as far at least as we can ascertain) on the waters of the ocean by producing tides; it is nevertheless highly *probable*, according to the observations of Messrs. Lambert, Toaldo, and Cotte, that in consequence of the lunar influence, great variations do take place in the atmosphere, and consequently in the weather. It would, indeed, extend too far the limits necessarily assigned to this article, to detail the ingenious reasonings of these eminent philosophers: but the following principles, extracted from their profound writings, will shew the grounds and reasons for their embracing the received notions on this interesting topic.

There are ten situations in every revolution of the moon in her orbit when she must particularly exert her influence on the atmosphere; and when, consequently, changes of the weather most readily take place. These are,

1. The *new*, and, 2. the *full* moon, when she exerts her influence in conjunction with, or in opposition to the sun.

3. & 4. The quadratures, or those aspects of the moon when she is 90 degrees distant from the sun; or when she is in the middle point of her orbit, between the points of conjunction and opposition, namely, in the first and third quarters.

5. The *perigee*, and, 6. the *apogee*, or those points of the moon's orbit, in which she is at the *least* and *greatest* distance from the earth.

The two passages of the moon over the equator, one of which M. Toaldo calls,

7. The moon's *ascending*, and the other, 8. the moon's *descending* equinox, or the two *lunistices*, as M. de la Lande terms them.

9. The *boreal lunistice*, when the moon approaches as near as she can in each lunation (or period between one new moon and another) to our zenith (that point in the horizon which is directly over our heads.)

10. The *austral lunistice*, when she is at the greatest distance from our zenith: for the action of the moon varies greatly according to her obliquity. With these ten points M. Toaldo compared a table of *forty-eight years'* observations; the result is, that the probabilities, that

the weather will change at a certain period of the moon, are in the following proportions:

New moon - - -	6 : 1
First quarter - - -	5 : 2
Full moon - - -	5 : 2
Last quarter - - -	5 : 4
Perigee - - -	7 : 1
Apogee - - -	4 : 1
Ascending equinox - -	13 : 4
Northern lunistice - -	11 : 4
Descending equinox - -	11 : 4
Southern lunistice - -	3 : 1

That is to say, a person may bet six to one, that the new moon will bring with it a change of weather. Each situation of the moon alters that state of the atmosphere which has been occasioned by the preceding one; and it seldom happens that any change in the weather takes place without a change in the lunar situations. These situations are combined, on account of the inequality of their revolutions, and the greatest effect is produced by the union of the syzigies*

* Syzigy, in astronomy, is a term equally used for the conjunction and opposition of a planet with the sun.

with the apsides.* The proportions of their power to produce variations are as follows:

New moon coinciding	with the perigee	33 : 1
Ditto	with the apogee	7 : 1
Full moon . . .	with the perigee	10 : 1
Ditto	with the apogee	8 : 1

The combination of these situations generally occasions storms and tempests; and this perturbing power will always have the greater effect, the nearer these combined situations are to the moon's passage over the equator, particularly in the months of March and September. At the new and full moons, in the months of March and September, and even at the solstices, especially the winter solstice, the atmosphere assumes a certain character, by which it is distinguished for three, and sometimes six months. The new moons which produce no change in the weather, are those that happen at a distance from the apsides.

* Apsides, in astronomy, are applied to two points in the orbits of planets, in which they are at the greatest and least distance from the sun or earth. The higher *apsis* is more particularly denominated aphelion, or apogee; the lower, perihelion, or perigee.

As it is perfectly true that each situation of the moon alters that state of the atmosphere which has been produced by another, it is however observed that many situations of the moon are favourable to good, and others to bad weather.

Those belonging to the latter class are: the perigee, new and full moon, passage of the equator, and the northern lunistice. Those belonging to the former are: the apogee, quadratures, and the southern lunistice. Changes of the weather seldom take place on the very days of the moon's situations, but either precede or follow them. It has been found by observation, that the changes effected by the lunar situations in the six winter months precede, and in the six summer months follow them.

Besides the lunar situations to which the above observations refer, attention must be paid also to the fourth day before new and full moon, which are called the *octants*. At these times the weather is inclined to changes; and it may be easily seen, that these will follow at the next lunar situation. Virgil calls this fourth day a very sure prophet. If on that day the horns of the moon are clear and well defined, good weather may be expected; but if they are dull, and not clearly marked on the edges, it is a sign that bad weather will ensue. When the weather remains unchanged on the fourth, fifth, and sixth day of the moon, we may conjecture that it will continue so till full moon, even sometimes till the next new moon; and in that case the lunar situations have only a very weak effect. Many observers of nature have also remarked, that the approach of the lunar situations is somewhat critical for the sick.

Conjectures on the periods of rain.

The rising and setting of the moon, as well as its superior and inferior passage of the meridian, may serve as a rule for foretelling the times of rain. These situations are called the moon's angles.

The times most exposed to rain are the rising

and setting; those most favourable to good weather, the passage of the meridian. It has been remarked that, during rainy days, bad weather is always a little interrupted about the time when the moon passes the meridian. We must, however, make an exception to this rule as often as the angle of the moon does not coincide with that of the sun. As these observations may be very easily made, by means of astronomical tables, in which the angles of the moon and sun are marked, they are exceedingly well calculated to prove the truth of this system. No one, for instance, will refuse assent to it, when the daily changes correspond with the angles of the moon; and when, independently of the effects of the moon's situation, the horizontal effect of the moon at rising and setting is different from that produced by its passage over the meridian.

It rains oftener in the day time than in the night, and oftener in the evening than in the morning.

Influence of the moon in regard to extraordinary years.

Bad years take place when the apsides of the moon fall in the four cardinal points of the zodiac. Their intervals, therefore, are as 4 to 5, 8 to 9, &c. or as the intervals of the passage of the apsides through the four cardinal points of the zodiac. Thus the year 1777 was, in general, a bad year; and in that year the apsides of the moon were in the equinoctial signs; and it is probable that the years in which the apsides fall in the signs Taurus, Leo, Virgo and Aquarius, will be good and moderate years, as the year 1776 really was; and in that year the apsides of the moon were in Taurus and Virgo.

Every eighteenth year must be similar. We, however, cannot depend upon a return altogether the same, on account of the three different revolutions of the moon; and therefore it may happen, that the epoch of this extraordinary year may be retarded a year or perhaps two. Though approximations only are here given, this does not prevent their being useful to farmers, if they only pay attention to circumstances. Besides, various exceptions must be made for different parts of the earth; and it is difficult to determine these beforehand, as what regards this system is applicable to the whole globe; but when the result of the system has been improved by local observations, the conjectures for each country will be attended with more certainty.

The fifty-fourth year must have a greater similarity to the first than all the rest; because, at this period, the situations of the moon, in regard to the sun and the earth, are again found in the same points.

The quantity of the rain which falls in nine successive years is almost equal to that which falls in the next following nine. But this is not the case when we compare in like manner the quantity of rain which falls in six, eight, or ten years.

Rules by Lord Bacon for prognosticating the weather, from the appearances of the moon.

1. If the new moon does not appear till the fourth day, it prognosticates a troubled air for the whole month.

2. If the moon, either at her first appearance or within a few days after, has her lower horn obscured and dusky, it denotes foul weather before the full; but, if she be discovered about the middle, storms are to be expected about the full; and, if her *upper* horn be affected, about the wane.

3. When on her fourth day the moon appears pure and spotless, her horns unblunted, and neither flat nor quite erect, but between both, it promises fair weather for the greatest part of the month.

4. An erect moon is generally threatening and unfavourable, but particularly denotes wind; though, if she appears with short and blunted horns, rain is rather to be expected.

We close these remarks on the probable influence of the moon on the weather, with the following Table; which has been ascribed to the illustrious astronomer, Dr. Herschel.* It is constructed upon a philosophical consideration of the attraction of the sun and moon in their several positions respecting the earth; and, confirmed by the experience of many years actual observation, will, without trouble, suggest to the observer what kind of weather will most probably follow the moon's entrance into any of her quarters; and that so near the truth, that in very few instances will it be found to fail.

* Europ. Mag. vol. 60, p. 24. The editor thinks it proper to give his authority for this table, which he has long and fruitlessly been searching in different philosophical publications.

NEW OR FULL MOON.	SUMMER.	WINTER.
If it be new or full moon, or the moon enters into the first or last quarters at the hour of 12....	Very Rainy...........	Snow and rain.
Or between the hours of		
2 and 4...........	Changeable	Fair and mild.
4....6...........	Fair	Fair.
6....8...........	Fair, if wind N. W..... Rainy, if S. or S. W....	Fair & frosty, if N. or N.E. Rainy, if S. or S. W.
8...10...........	Ditto	Ditto.
10...Midnight........	Fair	Fair and frosty.
Midnight....2...........	Ditto	Hard frost, unless Wind S. or S.W.
2....4...........	Cold, with freq. showers	Snow and stormy.
4....6...........	Rain	Ditto.
6....8...........	Wind and rain..........	Stormy.
8...10...........	Changeable	Cold, Rain if W. Snow if E.
10...Noon..........	Frequent showers	Cold, with high wind.

Hence, the nearer the time of the moon's entrance, at full and change or quarters, is to midnight (that is within two hours before and after midnight), the more fair the weather is in summer, but the nearer to noon the less fair. Also, the moon's entrance, at full, change, and quarters, during six of the afternoon hours, viz. from four to ten, may be followed by fair weather; but this is mostly dependent on the wind. The same entrance during all the hours after midnight, except the two first, is unfavourable to fair weather; the like, nearly, may be observed in winter.

Mr. Kirwan has lately endeavoured to discover probable rules for prognosticating the different seasons as far as respects Great Britain and Ireland, from tables of observations alone. On perusing and comparing a number of observations taken in England from 1677* to 1789, (a period of 112 years) he found:

1. That when there has been no storm before or after the vernal equinox, the ensuing summer is generally *dry*, at least five times in six.

2. That when a storm happens from an easterly point, either on the 19th, 20th, or 21st of May, the succeeding summer is generally *dry*, at least four times in five.

3. That when a storm arises on the 25th, 26th, or 27th of March (and not before) in any point, the succeeding summer is generally *dry*, four times in five.

4. If there be a storm at south-west, or west-south-west, on the 19th, 20th, 21st, or 22d of March, the succeeding summer is generally *wet*, five times in six.

In this country winters and springs, if dry, are most commonly cold; if moist, warm:—On the contrary, dry summers and autumns are usually hot, and moist summers cold; so that, if we know the moistness or dryness of a season, we

* Transactions of the Royal Irish Academy, vol. v. p. 20, &c.

can form a tolerably accurate judgment of its temperature. In this country also, Mr. Kirwan remarks, that it generally rains less in March than in November, in the proportion at a medium of 7 to 12. It generally rains less in April than October, in the proportion of 1 to 2, nearly at a medium. It generally rains less in May than September; the chances that it does so, are, at least, 4 to 3; but, when it rains plentifully in May (as 1.8 inches or more), it generally rains but little in September; and when it rains one inch, or less, in May, it rains plentifully in September.

From a table kept by Dr. Rutty, in Dublin, for *forty-one years*, Mr. Kirwan has endeavoured to calculate the probabilities of particular seasons being followed by others: although his rules chiefly relate to the climate of Ireland, yet as there exists but little difference between that island and Great Britain in the general appearance of the seasons, we shall mention his conclusions here,

In forty-one years there were

6 Wet springs, 22 dry, and 13 variable;
20 Wet summers, 16 dry, and 5 variable;
11 Wet autumns, 11 dry, and 19 variable.

A season, according to Mr. Kirwan, is accounted *wet*, when it contains two wet months. In general, the quantity of rain, which falls in dry seasons, is less than five inches, in wet seasons more; *variable* seasons are those, in which there falls between 30lbs. and 36lbs. a lb. being equal to ·157639 of an inch.

The order in which the different seasons followed each other was, as in the following table,

A *dry spring* has been followed by

a dry summer	11	times
a wet	8	
a variable	3	

A *wet spring* has been followed by

a dry summer	0	times
a wet	5	
a variable	1	

A *variable spring* has been followed by

a dry summer	5	times
a wet	7	
a variable	1	

A *dry summer* has been followed by

a dry autumn	5	times
a wet	5	
a variable	6	

A *wet summer* has been followed by

a dry autumn	5 times
a wet	3
a variable	12

A *variable summer* has been followed by

a dry autumn	1
a wet	3
a variable	1

Hence Mr. Kirwan deduced the probability of the kind of seasons that would succeed others, to be as follows.

In the beginning of any year,

I. The probability of a *dry spring* is	22-41
of a wet	6-41
of a variable	13-41
II. The probability of a *dry summer* is	16-41
of a wet	20-41
of a variable	5-41
III. The probability of a *dry autumn* is	11-41
of a wet	11-41
of a variable	19-41

IV. After a *dry spring*, the probability of

a dry summer is	1-22
a wet	8-22
a variable	3-22

V. After a *wet spring*, the probability of

a dry summer is	0
a wet	5.6
a variable	1.6

VI. After a *variable spring*, the probability of

a dry summer is	5.13
a wet	7.13
a variable	1.13

VII. After a *dry summer*, the probability of

a dry autumn is	5.16
a wet	5.16
a variable	6.16

VIII. After a *wet summer*, the probability of

a dry autumn is	5.20
a wet	3.20
a variable	12.20

IX. After a *variable summer*, the probability of

a dry autumn is	1.5
a wet	3.5
a variable	1.5

But the probability of the autumnal weather will be attained much more perfectly, by taking in the consideration of the preceding spring also; in order to which Mr. Kirwan observes that

A *dry spring* and *dry summer* were followed by a

dry autumn	3 times
wet	4
variable	4

A *dry spring* and *wet summer* were followed by a

dry autumn	2
wet	0
variable	6

A *wet spring* and *dry summer* were followed by a

dry autumn	0
wet	0
variable	0

A *wet spring* and *wet summer* were followed by a

dry autumn	2
wet	1
variable	1

A *wet spring* and *variable summer* were followed by a

dry autumn	1
wet	0
variable	0

A *dry spring* and *variable summer* were followed by a

dry autumn	0
wet	2
variable	1

A *variable spring* and *dry summer* were followed by a

dry autumn	2
wet	0
variable	1

A *variable spring* and *dry summer* were followed by a

dry autumn	2
wet	0
variable	2

A *variable spring* and *wet summer* were followed by a

dry autumn	1
wet	1
variable	5

A *variable spring* and *variable summer* were followed by a

dry autumn	0
wet	1
variable	0

X. Hence after *a dry spring* and dry summer, the probability of a

dry autumn is	3-11
wet	4-11
variable	4-11

XI. After a *dry spring* and *wet summer* the probability of a

dry autumn is	2-8
wet	0-11
variable	6-8

XII. After a *dry spring* and *variable summer*, the probability of a

dry autumn	0-0
wet	2-3
variable	1-2

XIII. After a *wet spring* and *dry summer*, the probability of a

dry autumn	0-41
wet	0-41
variable	0-41

XIV. After a *wet spring* and *wet summer*, the probability of a

dry autumn	2-5
wet	1-5
variable	2-5

XV. After a *wet spring* and *variable summer*, the probability of a

dry autumn	1-41
wet	0-41
variable	0-41

XVI. After a *variable spring* and a *dry summer*, the probability of a

dry autumn	2-4
wet	0-41
variable	2-4

XVII. After a *variable spring* and a *wet summer*, the probability of a

dry autumn	1-7
wet	1-7
variable	5-7

XVIII. After a *variable spring* and a *variable summer*, the probability of a

dry autumn	0-41
wet	0-41
variable	0-4

PART II.

OBSERVATIONS ON THE CHANGES OF THE WEATHER, INDICATED BY MEANS OF THE BAROMETER, AND OTHER PHILOSOPHICAL INSTRUMENTS.

In the preceding sections, it has been attempted to comprise the principal indications, afforded by the natural world, for ascertaining the various changes of the weather. Now, since every year, and the different seasons of each year, have a peculiar distinctive character, with regard to heat, cold, &c.; and further, since the quality of the seasons, has a very sensible effect on the productions of the earth, particularly in the drill and other systems of husbandry;—it is evidently of the greatest advantage to the farmer, to be able to foresee the nature of the ensuing changes, because he can thereby suit the culture of his ground, and his crops, to the weather expected.

SECTION I.

OF THE BAROMETER.

By means of that well known instrument, the Barometer, we are enabled to regain (in some degree at least) that fore-knowledge of the weather, which the ancients unquestionably did possess; though we know not the data on which they founded their conclusions. This instrument was first invented in the 17th century; and since that period it has received various important additions and improvements, which it would be foreign to this work to detail. We shall therefore annex such rules, as have hitherto been found most useful in ascertaining the changes of the weather, by means of the barometer.

1. The rising of the mercury presages, in general, fair weather; and its falling, foul weather, as rain, snow, high winds, and storms.

2. In very hot weather, especially if the wind is south, the sudden falling of the mercury foretels thunder.

3. In winter, the rising indicates frost; and in frosty weather, if the mercury falls three or four divisions, there will follow a thaw: but if it rises in a continued frost, snow may be expected.

4. When foul weather happens soon after the falling of the mercury, it will not be of long duration; nor are we to expect a continuance of fair weather, when it soon succeeds the rising of the quicksilver.

5. If, in foul weather, the mercury rises considerably, and continues rising for two or three days before the foul weather is over, a continuance of fair weather may be expected to follow.

6. In fair weather, when the mercury falls much and low, and continues falling for two or three days before rain comes, much wet must be expected, and probably high winds.

7. The unsettled motion of the mercury indicates changeable weather.

8. Respecting the words engraved on the register-plate of the barometer, it may be observed, that they cannot be strictly relied upon to correspond *exactly* with the state of the weather; though it will in general agree with them as to

the mercury rising and falling. The words deserve to be particularly noticed when the mercury removes from "*changeable*" upwards; as those on the lower part should be adverted to, when the mercury falls from "*changeable*" downwards. In other cases, they are of no use: for, as its *rising* in any part forebodes a tendency to *fair*, and its *falling* to *foul* weather, it follows that, though it descend in the tube from *settled* to *fair*, it may nevertheless be attended with a little rain; and when it rises from the words "*much rain*" to "*rain*," it shews only an inclination to become fair, though the wet weather may still continue in a less considerable degree than it was when the mercury began to rise. But if the mercury, after having fallen to "*much rain*," should ascend to "*changeable*," it foretels fair weather, though of a shorter continuance than if the mercury had risen still higher; and so, on the contrary, if the mercury stood at "*fair*" and descends to "*changeable*," it announces foul weather, though not of so long continuance as if it had fallen lower.

9. Persons who have occasion to travel much in the winter, and who are doubtful whether it will rain or not, may easily ascertain this point by the following observation.—A few hours before he departs, let the traveller notice the mercury in the upper part of the tube of the barometer: if rain is about to fall, it will be indented or concave; if *otherwise*, convex or protuberant.

The following remarks by a late eminent agriculturist* may serve more fully to elucidate the nature and uses of the barometer, to all who are engaged in agricultural pursuits. When (he observes) the character of the season is once ascertained, the returns of rain, or fair weather, may be judged of with some degree of certainty in some years, but scarcely guessed at in others, by means of the barometer; for, in general, we may expect, that when the mercury rises high, a few days of fair weather will follow. If the mercury falls again in two or three days, but soon rises high, without much rain, we may expect fair weather for several days; and in this case, the clear-

* Mr. Mills, in his "*Essay on the Weather*," p. 74.

est days are after the mercury begins to fall. In like manner, if the mercury falls very low, with much rain; rises soon, but falls again in a day or two, with rain; a continuance of bad weather may be feared. If the second fall does not bring much rain, but the mercury rises gradually pretty high, it prognosticates settled good weather of some continuance. When a heavy rain has fallen upon the mercury's sinking, and its continuing steadily low, the weather is sometimes fair, and promises well; but no prudent farmer should trust to such appearances. There is indeed a caution of this kind which the poorest may profit by. When the mercury rises high in the barometer, the moisture on the surface of the earth disappears; this, even though the sky be overcast, is a sure sign of fair weather; but if the earth continue moist, and water stands in shallow places, no trust should be put in the clearest sky, for it is in this case deceitful.

Towards the end of March, or more generally in the beginning of April, the barometer sinks very low, with bad weather; after which, it seldom falls lower than 29 degrees 5 minutes, till the latter end of September or October, when the quicksilver falls again low, with stormy winds, for then the winter constitution of the air takes place. From October to April, the great falls of the barometer are from 29 degrees 5 minutes, to 28 degrees 5 minutes, and sometimes lower; whereas during the summer constitution of the air, the quicksilver seldom falls lower than 29 degrees 5 minutes. It therefore follows that a fall of one tenth of an inch, during the summer, is as sure an indication of rain, as a fall of between two and three tenths is in the winter.

It must, however, be observed, that these heights of the barometer hold only in places nearly on a level with the sea; for experiments have taught us, that for every eighty feet of nearly perpendicular height that the barometer is placed above the level of the sea, the quicksilver sinks one tenth of an inch; observations alone therefore must determine the heights of the quicksilver, which in each place denote either fair or foul weather.

Very heavy thunder-storms happen, without sensibly affecting the barometer; and in this case the storm seldom reaches far. When a thunder-storm is attended with a fall of the barometer, its effect is much more extensive. When the quicksilver falls very low, and the weather continues mild and the wind moderate, there is at the same time a violent storm in some distant place: this accounts for a false prognostic, with which the barometer has often been unjustly charged. The effects which heat, cold and wind, severally produce on the glass, independently of the dry or humid state of the atmosphere, should likewise be considered.

From the preceding remarks and facts, it will be obvious to the reflecting reader that a barometer is almost as useful an appendage to the farmer as any other implement; for, as an intelligent writer in the Agricultural Magazine has observed, unless his operations are conducted with an attentive eye to the present or probably future state of the weather, as well as soil, the produce of his labours will either fall far short of his expectations, or (which to him is equally fatal) will suffer from ill-timed, though otherwise commendable exertions to house it.

VEGETABLE BAROMETERS.

The Cerea, or Night Barometer.

The CEREA is a native of Jamaica and Vera Cruz. It expands an exquisitely beautiful coral flower, and emits a highly fragrant odour, for a few hours in the night, and then closes, to open no more The flower is nearly a foot in diameter; the inside of the calyx, of a splendid yellow; and the numerous petals are of a pure white. It begins to open about seven or eight o'clock in the evening, and closes before sun-rise in the morning.

The flower of the DANDELION possesses very peculiar means of sheltering itself from the heat of the sun, as it closes entirely whenever the heat becomes excessive. It has been observed to open in summer, at half an hour after five in the morning, and to collect its petals towards the centre about nine o'clock.

CANINE BAROMETER.

The following anecdote of instinct in a dog (communicated by a correspondent) is too remarkable to be omitted: but, while we thus give it a place in the present work, we do not mean to affirm that the canine species may in every instance be considered as animal barometers.

A gentleman, some few years since, brought a pointer-dog from South Carolina, who was a remarkable prognosticator of bad weather.—"Whenever I observed him (says his master,) prick up his ears in a listening posture, scratching the deck, and rearing himself up, to look over to the windward, where he would eargerly snuff up the wind, though it was the finest weather imaginable, I was sure of a succeeding tempest; and this animal was grown so useful to us, that whenever we perceived the fit upon him, we immediately reefed our sails, and took in our spare canvas, to prepare for the worst.

SECTION II.

OF THE HYGROMETER.

The HYGROMETER is a contrivance, by which we are enabled to measure the degrees of dryness or moisture of the atmosphere. This instrument has long been neglected in meteorological observations: it is necessary to associate with it the THERMOMETER (which is noticed in a future page) and the BAROMETER, in order to be enabled to unravel the complication of different causes which influence the variations of the atmosphere; and it is only by a long series of observations, made by these various instruments, together with all the indications deduced from the state of the heavens, that we can obtain such data as will enable us to prognosticate (with great probability) the temporary changes, and to arrive at a plausible theory upon this interesting subject.

There are various sorts of hygrometers: for, whatever body either swells or shrinks by moisture or dryness, may be formed into a hygrometer. Such are most kinds of wood, particularly white wood, as poplar, birch, deal, &c. And on this principle it is, that wedges of well dried wood are employed for cleaving or raising rocks or stones; for, in proportion as the moisture of dew, rain, or water, applied to them, enters into them, they swell and overcome an inconceivable resistance. Ropes or strings made of hemp, flax, or any other vegetable substance, become also hygrometers. This is well known to sailors, who, according to the dryness or moisture of the air, find the shrouds of their vessels slack or tightened, so as, in the latter case, to be in danger of breaking.

Stretch a cord or fiddle-string, fastened at one end over a pulley, and to the other end tie a weight; this will rise or fall as the air becomes dry or moist, and consequently be an hygrometer.

Animal substances, as catgut, whalebone, &c. twisted and dried, answer the same purposes, as performers on stringed instruments often find to their cost, when the too great moisture of the air breaks their strings. The Dutch toys, known by the name of *weather-houses*, are very good hygrometers for common purposes, and are formed on this principle. The contraction of the string, by moisture in the atmosphere, forces the male figure out of the door at the approach of *bad* weather; and, as this gradually becomes dry, the string resumes its natural length, and forces the female out of door, at the approach of *good* weather.

A great misfortune, however, which attends the use of all these substances is, that by use they become sensibly less and less accurate, so as at length not to undergo any visible alteration from the different states of the air, in regard to dryness or moisture. On this account a sponge may be preferred, as being less liable to be so changed. To prepare the sponge, first wash it in water, and when dry, wash it again in water wherein sal ammoniac, or salt of tartar, has been dissolved; and let it dry again. Now, if the air becomes moist, the sponge will grow heavier; and if dry, it will become lighter.

Oil of anise-seeds, with proportions of oil of almonds or Florence oil, might serve to measure degrees of heat or cold, and other appearances of the weather. Oil of vitriol is found to grow sensibly lighter or heavier in proportion to the lesser or greater quantity of moisture it imbibes from the air. The alteration is so great, that it has been known to change its weight from three drams to nine. The other acid oils, or, as they are usually called, spirits, or oil of tartar *per deliquium*, may be substituted for the oil of vitriol.

In order to make an hygrometer with those bodies which acquire or lose weight in the air, place such a substance in a scale on the end of a steel-yard, with a counterpoise which shall keep it in equilibrio in fair weather; the other end of the steel-yard, rising or falling, and pointing to a graduated index, will shew the changes.

If a line be made of good well-dried whipcord, and a plummet be affixed to the end of it, and the whole be hung against a wainscot, and a line be drawn under it, exactly where the plummet reaches, in very moderate weather it will be found to rise above such line, and to sink below it when the weather is likely to become fair.

The awn of barley also furnishes a simple but

efficacious hygrometer. It is furnished with stiff points, which, like the teeth of a saw, are all turned towards the point of it; as this long awn lies upon the ground, it extends itself in the moist air of night, and pushes forward the barley-corn, which it adheres to; in the day it shortens as it dries; and as these points prevent it from receding, it draws up its pointed end; and thus, creeping like a worm, will travel many feet from the parent stem. That very ingenious mechanic philosopher, Mr. Edgeworth, once made on this principle a wooden automaton. Its back consisted of soft fir-wood, about an inch square, and four feet long, made of pieces cut the cross-way, in respect to the fibres of the wood, and glued together: it had two feet before, and two behind, which supported the back horizontally; but were placed with their extremities (which were armed with sharp points of iron) bending backwards. Hence, in moist weather, the back lengthened, and the two foremost feet were pushed forwards; in dry weather, the hind feet were drawn after, as the obliquity of the points of the feet prevented it from receding. And thus, in a month or two, it walked across the room which it inhabited. Might not this machine be applied as an hygrometer to some meteorological purpose?

A very simple hygrometer, Mr. Marshall states, may be formed by means of "a flaxen line, (large well-manufactured whip-cord) five feet long; and having a graduated scale fixed to an index, moving on a fulcrum. The length of the index, from the fulcrum to the point, should be ten inches; that of the lever, from the fulcrum to the middle of the eye, to which the cord is fixed, two and a half." He adds, that "the principle on which this hygrometer acts is obvious. The air becoming moist, the cord imbibes its moisture; the line, in consequence, is shortened, and the index rises. On the contrary, the air becoming dry, the cord discharges its moisture,—lengthens,—and the index falls. It may be true," he says, "that no two hygrometers will keep pace with each other sufficiently to satisfy the curious. He will venture to say, however, from seven months' close attention, that two hygrometers, on this simple construction, have coincided sufficiently for the uses of agriculture. It is true," he adds, "they diminished in the degree of action; but as the scale may be readily diminished in extent, and as a fresh line may be so cheaply and so readily supplied, this is not a valid objection." It is remarked, that "this diminution, in the degree of action, depends considerably on the construction; the propriety, or rather delicacy, of which, rests, almost solely, on this point: the weight of the index should be so proportioned to the weight of the lever and cord, that the cord may be kept perfectly straight, without being too much stretched. He made one with a long heavy index; and, in order to gain a more extensive scale, with a short lever; but, even when it was first put up, it could barely act; and, in a few weeks, it flagged, and was not able to raise the index, though the air was uncommonly moist. He therefore made another, with the same length, both of index and lever, but with a lighter index, and a heavier lever, so as to gain the proportion above-mentioned; and it has acted exceedingly well." He thinks that no farmer, "who wishes to profit by the hygrometer, should have less than two. Three or four would be more advisable. They would then assist in correcting each other; and, in case of renewal or alteration, there would be no danger of losing the state of the atmosphere; which, if only one is kept, must necessarily be the case. The principle on which this hygrometer is formed, is not, he says, confined to a small cord, and an index of ten inches long: it may be extended to a rope, of any length or thickness, and to an index and scale, of almost any dimensions and extent." But one, or more, on a portable construction, might, he thinks, be found useful. An axe is the form he has thought of; the edge, graduated, will constitute the scale; and the handle will receive the cord: this may be hung up, in the shade, exposed to the action of the air; or, by means of a spike in the end of the handle, it may be placed in the open field. By placing it on fallow ground, it may be actuated by the perspiration of the earth; among vegetables, by vegetable perspiration. By the means of one, or, more probably, by the means of several placed at varied heights, the different degrees of moisture at different altitudes may be ascertained, &c. In fact, he considers the hygrometer, whether it is a prognostic of the weather or not, as a most valuable oracle to the farmer.

How valuable an oracle this instrument *may* prove, the reader may easily conceive from the following extract of Mr. Marshall's "Minutes of Agriculture."

"Yesterday morning, while the hygrometer stood at two degrees moist, the peas were by no means fit for carrying; the haulm was green, and the peas were soft. About ten o'clock, the hygrometer fell to one degree dry; before one, the peas were in good order, I went up into the field, merely on the word of the hygrometer, and found them fit to be carried."

If, however, the observer be desirous of instituting very accurate experiments, it will be advisable to procure the whalebone hygrometer, originally invented by M. De Luc, which is esteemed one of the best now in use.

SECTION III.

THE RAIN-GAUGE.

The Rain-Gauge (also termed a *Pluviometer)* is a machine for measuring the quantity of rain that falls. One of the best constructed rain-gauges consists of a hollow cylinder, having within it a cork-ball attached to a wooden stem, which passes through a small opening at the top, on which is placed a large funnel. When this instrument is placed in the open air in a free place, the rain that falls within the circumference of the funnel will run down into the tube and cause the cork to float: and the quantity of water in the tube may be seen by the height to which the stem of the float is raised. The stem of the float is so graduated, as to shew by its divisions the number of perpendicular inches of water which fell on the surface of the earth since the last observation. After every observation the cylinder must be emptied.

Another very simple rain-gauge may be formed of a copper funnel, the area of whose opening is exactly ten square inches. Let this funnel be fixed in a bottle, and the quantity of rain caught is ascertained by multiplying the weight in ounces by .173, which gives the depth in inches and parts of an inch. In fixing these gauges, care must be taken, that the rain may have free access to them; hence the tops of buildings (according to Mr. Nicholson*) are usually the best places, though some conceive that the nearer the rain-gauge is placed to the ground, the more rain it will collect.

In order to compare the quantities of rain collected in pluviometers at different places, the instruments should be fixed at the same heights above the ground in both places; because, at different heights, the quantities are always different, even at the same place.

* "British Encyclopedia," article *Rain-Gauge*.

SECTION IV.

OF THE THERMOMETER.

Besides a barometer for measuring the *weight* of the atmosphere, a thermometer is equally necessary, in order to shew the variations in the *temperature* of the weather: for every change of the weather is attended with a change in the temperature of the air, which a thermometer placed in the open air will point out, sometimes before any alteration is perceived in the barometer.

The knowledge of the exact degree of cold in the winter is of consequence to the farmer: for it has been observed, that when the frost is so keen that the thermometer sinks fourteen degrees on Fahrenheit's scale, most succulent vegetables are thereby destroyed, such as almost all the cabbage or kale tribe, turnips, &c.; for their juices being then frozen hard, their vessels are thereby torn asunder or split, so that when the thaw comes on, the whole substance, for instance of turnips and apples, runs into a putrid mass. In this case the most likely way to prevent their being lost, is to immerse what is so frozen in cold water, till the frost is extracted by the water: the loss is thereby delayed a little, for what is not used very speedily will soon putrify, notwithstanding this care. The knowledge of this consequence of so severe a frost, may however suggest to the farmer some method of repairing the loss he expects. Time may point out other useful observations, which may arise from the knowledge of what may be discovered from the changes in the thermometer.

The thermometer was invented in the seventeenth century; and from its extensive utility in the arts, manufactures, and domestic life, the honour of its invention has been attributed to various eminent men. Like the barometer, it has also received various improvements; but that chiefly used in this country is Fahrenheit's, though in France and some parts of the continent Reaumur's (which is less accurate) is employed.

The scale affixed to Fahrenheit's thermometer is divided into degrees or equal parts; its freezing point is 32 degrees above 0 (or zero, as it is called by philosophers,) and boiling water 212 degrees. As, however, Reaumur's scale is sometimes (though rarely) used, it may be proper to add, that its freezing point is 0, and boiling water at 80 degrees.

From some very accurate tables constructed by the late Mr. Kirwan, it appears that January is the coldest month in every latitude; and that July is the warmest month in all latitudes above 48 degrees: in lower latitudes, August is generally the warmest. The difference between the hottest and coldest months increases in proportion to the distance from the equator. Every habitable latitude, he further remarks, enjoys a mean heat of 60 degrees for at least two months; which heat is necessary for the production of corn.

But the *only method*, by which the changes of the weather can be traced with precision, is, to keep regular registers of the weather, and to mark every apperance in the heavens or on the earth, which may tend to point out the approaching seasons. This point cannot be urged too strongly on the attention of the intelligent agriculturist: for, as the pursuits of a farmer necessarily require him to be much in the open air, this office would become both regular and easy to him: and his progress in fixing FACTS, and in drawing judicious conclusions from them, would probably be more speedy and successful than he might otherwise expect, and would enable him, *profitably*, to regulate the management of his crops.

This important object might in all probability be more effectually obtained, if, together with the usual registers of the weather, observations were made on the winds in many parts of the earth. For this purpose the three following instruments have been suggested: they may be constructed at no great expence, and thus some useful information might be acquired.

1. To mark the hour when the wind changes from north-east to south-west, and the contrary.—This might be managed by making a communication from the vane of a weathercock

to a clock, in such a manner, that if the vane should revolve quite round, a tooth of its revolving axis should stop the clock, or put back a small bolt on the edge a wheel, revolving once in twenty-four hours.

2. To discover whether in a year more air is passed from north to south, or the contrary.— This might be effected by placing a windmill-sail of copper, about nine inches diameter, in a hollow cylinder, about six inches long, open at both ends, and fixed on an eminent situation, exactly north and south. Thence only a part of the north-east and south-west currents would affect the sail so as to turn it; and if its revolutions were counted by an adapted machinery, as the sail would turn one way with the north currents of the air, and the contrary one with the south currents, the advance of the counting finger either way would shew which wind had prevailed most at the end of the year.

3. To discover the rolling cylinder of air,— the vane of a weathercock might be so suspended as to dip or rise vertically, as well as to have its horizontal rotation.

PART III.

THE SHEPHERD OF BANBURY'S RULES, BY WHICH TO JUDGE OF THE CHANGES OF THE WEATHER, (GROUNDED ON FORTY YEARS EXPERIENCE), METHODIZED AND ARRANGED UNDER DISTINCT HEADS.

SECTION I.

PROGNOSTICS OF THE WEATHER, TAKEN FROM THE SUN, MOON, AND STARS.

I. SUN. *If the sun rise red and fiery.*— Wind and rain.

Remarks.

The old English rule published in our first almanacks agrees exactly with our author's observation.

If red the sun begins his race,
Be sure that rain will fall apace.

The shepherd begins with observations arising form the different appearances of the sun. These rules may be extended to all the heavenly bodies: for, as their rays pass through the atmosphere, the vapours in the air have the same effect on each. Thus,

The *rain-bow* shews us that the rays of light admit of different degrees of refraction, and that according to those different degrees of refraction, they appear of different colours. A clear unclouded sky teaches us, that while the vapours are equally dispersed in the atmosphere, the rays reach us without undergoing a change, or variety of colours. It is known to those conversant in experimental philosophy, that this refraction of the rays of light arises from a difference in the density of the medium through which the rays pass. It seems probable, that while the watery vapour in the air is divided into its minutest particles, it perhaps only reflects the rays of light, but does not refract them till collected into the form of water, as into clouds, rain, &c. When the farmer therefore sees the sun or moon rise or set red and fiery, or sees the clouds and horizon of that colour, he may expect wind and rain, owing to the unequal distribution of the vapours, or to their being already collected into watery globules by some preceding cause.

The *circle* which frequently appears about the *moon*, and sometimes about the *sun*, as also the *mock-suns* and *moons*, proceeding from the great quantity of watery vapour loading the lower air, likewise presage *rain* or *wind*, and often both.

II. *If cloudy, and the clouds soon decrease.*—
Certain fair weather.

Remarks.

This is in consequence of vapours being more equally distributed in the atmosphere; which equal distribution is also promoted by the warmth of the rising sun. Hence we may account for an observation adopted into all languages,

The evening *red*, and the morning *grey*, is a sign of a *fair* day.

For if the abundance of vapour denoted by the *red* evening sky falls down in dew, or is otherwise so equally dispersed in the air, that the morning shall appear *grey*, we may promise ourselves a *fair* day, from that equal state of the atmosphere.

If, in the morning, some parts of the sky appear *green* between the clouds, while the sky is *blue* above, *stormy* weather is at hand.

SECTION II.

PROGNOSTICS OF THE WEATHER TAKEN FROM THE CLOUDS.

III. *Clouds small and round, like a dapple-grey, with a north-wind*—Fair weatherfor two or three days.

Remarks.

This is differently expressed by other authors. Thus Lord Bacon observes, that if clouds appear white, and drive to the north-west, it is a sign of several days fair weather.

Our old English almanacks have a maxim to this purpose:

> If woolly fleeces spread the heavenly way,
> Be sure no rain disturbs the summer day.

And Pliny to the same purpose.

Si sol oriens cingetur orbe, et postea totus defluxerit æqualiter, serenitatem dabit.

That is,

If the rising sun be encompassed with an iris or circle of white clouds, and both of them fly away, this is a sign of fair weather.

There is another English proverb worth remembering,

> In the decay of the moon,
> A cloudy morning bodes a fair afternoon.

This rule, however, seems to contradict an observation made by Mr. Worlidge, viz. that "In a fair day, if the sky seem to be dappled with clouds, (which is usually termed a mackarel sky), it generally predicts *rain*." This is confirmed by another observer of nature, who has constantly found, that, in dry weather, so soon as clouds appear at a great height striped like the feathers in the breast of a hawk, *rain* may be expected in a day or so.

Mr. Worlidge proceeds thus. " In a clear evening, certain small black clouds appearing, are undoubted signs of *rain* to follow; or if black or blue clouds appear near the sun at any time of the day, or near the moon by night, *rain* usually follows.

" If small waterish clouds appear on the tops of hills, *rain* follows; as they observe in Cornwall, that

" *When Hengston is wrapped with a cloud, a shower follows soon after.*

" The like they observe of Rosemary-topping, in Yorkshire, and many other places in England.

" If clouds grow or appear suddenly, the air otherwise free from clouds, it signifies *tempests* at hand, especially if they appear to the south or west."

If many clouds, like fleeces of wool, are scattered from the east, they foretel *rain* within three days.

When clouds settle on the tops of mountains, they indicate hard weather; and when the tops of mountains are clear, it is a sign of fair weather.

IV. *If small Clouds increase*—Much Rain.

V. *If large Clouds decrease*—Fair Weather.

VI. *In Summer or Harvest, when the Wind has been South two or three Days, and it grows very hot, and you see Clouds rise with great white Tops like Towers, as if one were upon the Top of another, and joined together with black on the nether Side*—There will be Thunder and Rain suddenly.

VII. *If two such Clouds arise, one on either hand*—It is Time to make haste to shelter.

Remarks.

The two last of these rules may be illustrated by the following extract from Boerhaave's System of Chemistry.

"If a large white, what may be supposed a frozen cloud, be opposed to the sun, the rays reflected by the side next the sun must rarefy or heat the air between it and the sun, while at the same time, allowing that the cloud is not transparent, the cold will be great in the part turned from the sun, and the air so much the denser: whence must arise a violent motion of the cloud, which will be the more rapid, in proportion as the sun's heat is the greater on one side, and the cold is the keener on the other side. If a few such clouds are so disposed, that their joint effects meet in one place, which may often be the case, it is easy to conceive that a very great heat must suddenly arise in such a place, and the air be as greatly expanded therein. On a change of the situation of the clouds, and a consequent dissipation of the rays of the sun, the heat ceases, and the cold air, snow, hail, rain, or other substances near at hand, will rush violently into the spaces so heated; whence most stupendous and destructive effects may be produced. Hence it will not be surprising, that a small cloud appearing in a clear sky, in a hot climate, still increasing till it reaches the earth, produces those direful effects travellers acquaint us they meet with in certain latitudes: and thus, even in our northern climate, small white clouds are sometimes seen at a good height, especially after a drought or calm, continually increasing, and as they increase, turning less and less white, till at length they burst down in heavy showers, which falling in large drops, shew that they come from a considerable height, and that they had probably been hail. As the air admits of greater rarefaction than water, the watery vapour must consequently precipitate out of the heated rarefied air. From this cause the inequality of rain in such showers may proceed."

VIII. *If you see a Cloud rise against the Wind or side Wind, when that Cloud comes up to you*—The Wind will blow the same Way that the Cloud came. *And the same Rule holds of a clear Place, when all the Sky is equally thick, except one Edge.*

Remarks.

As wind is nothing more than air in motion, the effects of it first discover themselves above, and actually drive such clouds before them: this was long ago observed by Pliny. When clouds, says he, float about in a serene sky, from whatever quarter they come, you may expect winds. If they are collected together in one place, they will be dispersed by the approach of the sun. If these clouds come from the north-east, they denote winds; if from the south, great rains. But let them come from what quarter they will, if you see them driving thus about sun-set, they are sure signs of an approaching tempest.

If the clouds look dusky, or of a tarnish silver-colour, and move very slowly, it is a sign of hail. But to speak more plainly, those very clouds are laden with hail, which, if there be a mixture of blue in the clouds, will be small, but if very yellow, large. Small scattering clouds that fly very high, especially from the south-west, denote whirlwinds. The shooting of falling stars through them, is a sign of thunder. We meet with many observations of this sort in our old writers on husbandry, and we have abundance of proverbs relating to this subject which are worth observing, and the rather, because most of them are not peculiar to our language only, but common to us with many of our neighbours. Lord Bacon has very judiciously remarked, that proverbs are the philosophy of the common people, that is to say, they are trite remarks founded in truth, and fitted for memory. Some of them, it must be confessed, seem either false, or of no great consequence; but it is highly probable in such cases, that by various accidents we have lost their true meaning, or else, that in length of time they have been altered and corrupted, till they have little or no meaning at all.

SECTION III.

PROGNOSTICS OF THE WEATHER TAKEN FROM MIST.

IX. *If Mists rise in low Grounds, and soon vanish*—Fair Weather.

Remarks.

This is a certain sign, and well expressed; and its correctness will be more fully evinced when the nature of mists is considered.

Mists are gross vapours, which while they float near the earth are styled mists, but when they ascend into the air, are called clouds. If, therefore, rising out of low ground, they are driven along the plain, and are soon lost to the sight, it must arise from some of these causes, viz. That there is sufficient air abroad to divide and resolve them, or the heat of the sun has been strong enough to exhale them, that is, to rarefy them, so as to render them lighter than the air through which they were to pass. Whichever way this happens, the maxim remains unimpeached.

X. *If Mists rise to the Hill-tops*—Rain in a Day or two.

Remarks.

When mists are very heavy in themselves, and rise only by the action of that protrusive force, exerted by the subterranean fire, they can rise no higher than where the gravitation becomes superior to that protrusive force, for then they descend again by their own weight, and this occasions the appearance mentioned in the observation of their hanging upon hill-tops, where they are very soon condensed, and fall down in rain.

Formerly there was a very idle and ill-grounded distinction between moist and dry exhalations, whereas, in truth, all exhalations are moist, or in other words, are watery streams thrown off by bodies respectively dry; and the former distinction was invented only to solve these phenomena of which we have been speaking, that is, the mist rising and dispersing without rain, and the mist condensed and resolved into rain.

XI. *A general Mist before the Sun rises, near the full Moon*—Fair Weather.

Remarks.

This is a general and a very extensive observation, which enables us to judge of the weather for about a fortnight, and there is very great reason to believe that it will very rarely deceive us.

Mists are observed to happen when the mercury in a barometer is either very low or very high. They happen when it is high after the region of the air has continued calm a good while, and in the mean time a great abundance of vapours and exhalations have been accumulated, making the air dark by their quantity, and the disorderly disposition of their parts. They happen when the mercury is low, sometimes because the rarity of the air renders it unable to sustain the vapours, which therefore descend and fall through it.

XII. *If Mists in the New Moon*—Rain in the Old.

Remarks.

When exhalations rise copiously from the earth into the region of the air, and the air itself is in a proper disposition, they ascend to a great height, and continue a long time before they are condensed; which accounts very clearly and philosophically for the interval of fair weather between the rising of these mists, and their falling down again in showers. Their ascending about sun-rise is a proof that the air is thin, but at the same time of a force sufficient to sustain them, since if the mists were not specifically lighter than the air itself they could not ascend.

When the moon is at the full, and such exhalations rise copiously, the time necessary for them to float in the atmosphere, before they are condensed into clouds and rain, generally extends beyond that moon, and therefore the present observation directs us to expect fair weather.

XIII. *If Mists in the Old*—Rain in the New Moon.

Remarks.

It is an observation, applicable to every climate, that great changes of the weather happen at the changes of the moon. It follows that this is the season when the exhalations, that ascend so copiously at sun-rise, are condensed; and consequently at this season we must expect rain. If therefore exhalations rise in the new moon, it indicates that the air is in a fit disposition to support them for some time: consequently, we may expect them to continue floating till the next regular change of weather, that is, till the old of the moon, or rather till towards the next change. The observation, therefore, is very properly and cautiously worded, directing us to expect rain IN the old and IN the new, and not AT the old or new; because experience shews that these changes of the weather happen not exactly at the change of the moon, but a day or two before or after. Several instances of this occur in Capt. Dampier's "*History of Winds and Storms at Sea.*"

SECTION IV.

PROGNOSTICS OF THE WEATHER, TAKEN FROM RAIN.

XIV. *Sudden Rains never* last long: *But when the Air grows thick by Degrees, and the Sun, Moon and Stars shine dimmer and dimmer, then it is like* to rain six Hours usually.

Remarks.

A sudden rarefaction of the lower air, or perhaps more frequently a cold cloud descending from above, or cold wind descending from above and condensing the invisible vapours so as to form a cloud, are the most frequent causes of sudden rain. The rain, therefore, ceases as soon as an equal temperature is restored to the atmosphere: but if the vapours are collected in the manner described in the latter part of this rule, it is no wonder that the rain continues longer.

Mountainous countries, it is observed, have most rain, and the reason seems to be the winds driving the clouds against the rocks and hills, and thereby compressing them in such a manner, that they are immediately dissolved, and fall as it were at once. Thus, in Lancashire there falls twice as much rain as in Essex, and from the same cause, probably, in the ocean over-against the mountainous coast of Guinea, showers sometimes fall, as it were, by pails full.

This observation of our shepherd is very just and reasonable, and will rarely fail such as observe it. The dimness of the stars and other heavenly bodies, is one of the surest signs of very rainy weather. It is likewise to be observed, that when the stars look bigger than usual, and are pale and dull and without rays, this undoubtedly indicates that the clouds are condensing into rain, which will very soon fall; and it has been obseved, that when the air grows thick by degrees, and the light of the sun lessens so as not to be discerned at all, and again when the moon and stars have the same appearances, a continued rain for at least six hours is sure to follow.

In order to have the most certain information in such cases, it is best to have recourse to a variety of signs: for not only do the clouds and sky, or the sun, moon, and stars, give us previous notice of rainy weather, but almost every thing in the creation, and vegetables particularly. For instance, the pimpernel, which is a very common flower, shuts itself up extremely close against rainy weather. In like manner, the trefoil swells in the stalks against rain, so that it stands up very stiff, but the leaves droop, and hang down. Even the most solid bodies are affected by this change of the atmosphere, for stones seem to sweat, and wood swells, the air driving the moist particles with which it is filled into the pores of dry wood especially, make it swell prodigiously; and this is the reason why doors and windows are hard to shut in rainy weather.

This is so true, that a method has been invented of dividing mill-stones by the mere force of the air, which is done in the following manner. They divide a block of this kind of stone as big as a large rolling stone, into as many parts as they design to make mill-stones, and in the circles where this block is to be divided, they pierce several holes, which they fill with aloes-wood dried in an oven, and expose the stone to the air in moist weather; when the wood swells to such a degree as to split the stone as effectually, as if it was by iron wedges driven by sledge hammers.

XV. *If it begin to rain from the South, with a high Wind for two or three Hours, and the Wind falls, but the Rain continues,* it is like to rain twelve Hours or more, and does usually rain till a strong North Wind clears the Air. *These long Rains seldom hold above twelve Hours, or happen above* once a Year.

Remarks.

In the state of the air described in this rule, the mercury in the barometer will always be found low, which indicates that the atmosphere is light. The rain, therefore, continues to fall, till a cooler and denser air from the north enables the atmosphere to support the vapours.

The duration of rain in an *inland county*, like Oxfordshire (where the shepherd resided) may not exceed *twelve hours:* but it is questionable whether this will hold as a general rule, either as to its duration or its frequency, in all places; for, near the sea, rains often happen, and continue for a whole day.

XVI. *If it begins to rain an Hour or two before Sun-rising*, it is likely to be fair before Noon, and to continue so that day; *but if the Rain begins an Hour or two after Sun-rising*, it is likely to rain all that Day, except the Rain-bow be seen before it rains.

Remarks.

This is a short clear and easy observation, requiring but few remarks: a few hints, however, may not be irrelevant on the formation of the rain-bow.

The rain-bow, then, is a circular image of the sun, variously coloured, and is thus produced: The solar rays, entering the drops of falling rain, are refracted to their further surfaces, and thence, by one or more reflections, transmitted to the eye: at their emergence from the drop, as well as at their entrance, they suffer a re-fraction, by which the rays are separated into their different colours, which consequently are most beautifully exhibited to an eye properly placed to receive them.

Sometimes (though rarely) two, and even three, rain-bows are seen: the colours in the bow are thus disposed, viz. violet, purple, blue, green, yellow, orange, red. After a long drought, the bow is a certain sign of rain; if after much wet, fair weather.—If the green be large and bright, it is a sign of rain, but if the red be the strongest colour, then it denotes wind and rain together.—If the bow breaks up all at once, there will follow serene and settled weather.—If the bow be seen in the morning, small rain will follow; if at noon, settled and heavy rains; if at night, fair weather. The appearance of two or three rain-bows shews fair weather for the present, but settled and heavy rains in two or three days time.

LUNAR RAIN-BOWS.—The moon sometimes exhibits the phenomenon of a rain-bow by the refraction of her rays in drops of rain in the night-time. Lunar rain-bows very seldom present themselves to our observation; they are extremely beautiful, though much less than those that appear in the day-time, and a yellow or rather a straw-colour chiefly prevails. As they are of such rare occurrence, they cannot well be reckoned among the signs of weather; consequently no probable rules for ascertaining the weather can be deduced from the appearance of such rain-bows.

SECTION V.

PROGNOSTICS OF THE WEATHER TAKEN FROM THE WINDS.

When the atmosphere is of the same weight and density over a considerable extent of the surface of the earth, there a calm will obtain: but if this equipoise is taken off, a stream of air, or wind, is produced, stronger or weaker in proportion to the alteration made in the state of the atmosphere. There are divers causes which make these alterations in the equipoise of the atmosphere, such as rarefactions or condensations in one part more than in another; vapours rising from the earth or sea, pressure of the clouds, &c. It would be foreign from the nature of this work to enter into a disquisition concerning the causes of the winds in general; we shall therefore refer the curious to Lord Bacon, Mr. Bohun, Dr. Halley, Dr. Franklin, and others who have written more fully on this subject, and confine our attention chiefly to the winds so far as relates to this island.

Three causes may be assigned for the stated winds in this island. The first of these stated winds is the *westerly*, which so frequently obtains every where beyond the limits of the trade-wind, and has been most judiciously accounted for by Dr. Franklin. This general westerly wind is found to blow mostly from the north-west in the ocean, and where other causes do not intervene. Lord Bacon mentions the other two causes, as having been long observed, *viz.* that winds blow most frequently from the sea; and next, that where there are high mountains covered with snow, stated winds blow from that quarter at the time the snow dissolves.

Lord Bacon imputes the frequency of the winds from the sea to the copious ascent of watery vapour from it; and as signs that such vapours do ascend from it, he observes, that "the sea and lakes sometimes swell very considerably, though no winds are found to blow, which he remarks is probably occasioned by the warm vapour rising out of the earth under the water. At such times a kind of murmuring noise is heard, the sounding of the shore is heard to a greater distance than usual, and sometimes a froth or watery bubbles are seen on the sea, whilst it is flat and calm. Hence miners foretel storms, by the muddiness of the water, or by the fumes which rise in mines, before any signs appear above ground." Mr. Bohun relates, that " in Cornwall they have so sure prognostics of storms at sea, from their mines, that the fishermen never presume to remain out, when the signal is given by the eruption of certain meteors, which immediately presage a tempest. In St. Owen's bay in the isle of Jersey," continues he, " the sea is often strangely disturbed before the western storms, even when the air is very calm; and though no wind be stirring, yet the roaring of the waves may be heard, not only over the whole isle, but into France about thirty miles distant, which is the certain prognostic of an ensuing tempest."

This agitation of the sea, and noise of the water, may be occasioned by a storm in the Atlantic Ocean, with the wind at west; for as the storm proceeds eastward, the waves raised by it will greatly outgo the wind, and thereby reach the eastern coast some hours before the wind arrives there. It is probable, that if any storms arise from vapours ascending thus from the earth under the sea, they are only such as are very violent: for that power which the air has of taking up water, will supply sufficient to occasion the winds so frequent from the sea, and is perhaps their most general cause.

Wind is air in motion, excited by various causes. The sun, by concurrent circumstances in land, water, and vapour, lightens and disperses the air from one place, and at one time, more than at another. Inflammable exhalations, and other explosions, shall warm and thin the air in particular places. A cloud or portion of vapour full of electrical matter, passing over a cloud or region of land more destitute of electrical matter, will shed streams of fire upon the less electric body, and thereby excite violent motions, &c. Now, wherever the air is thrown into a state of rarefaction, there a vacuity is produced, and the adjacent air flows as water to the breach of a dam, and the flood is either violent or not, as the space through which it passes is shaped; lasting as the quantity of fluid set in motion, and as the extent of the vacuity is to be replenished. If the vacuity be spacious, the flow will be plentiful (obstructions in the way being allowed for); if the channel through which the influx runs be long, narrow, and funnel-like, the velocity will be great, and *vice versa;* but if a large quantity of condensed air should at this time press forward towards this large vacuity, the motion of the air will be impetuous, or what we call a *storm*. If, on the other hand, the rarefactions in particular districts be gentle, and there is room for denser air to succeed without violence, the motion also is gentle; and where no extraordinary rarefactions are produced, and the vapours are equally dispersed, a calm ensues.

If rarefying vapours assume the shape of an oblate disc, over-spreading as a canopy a wide extent, the weight and continuity of the incumbent air is in this district, for a time, and to a certain degree, suspended; the mercury sinks in the barometer, and at the same time the current of the air above this disc shall go one way, towards any vacuity, which shall create a fresh tendency, and the under-current of air, influenced by another rarefaction, shall go in a different, perhaps opposite direction; there being no communication between the currents above and below the disc of vapours, sufficient to determine them to one point. Thus again, by the fall or even recess of a great body of vapours in one place out of our sight, the air over our heads being condensed, and keeping the mercury high, extends itself into the vacuity, the wind blows, and the mercury falls in a serene sky, to our surprise. By the rising of a like body of vapours, and accumulating the air of our horizon, the mercury rises in a cloudy and even rainy sky. When the wind is violent, the perpendicular pressure of the air is much lessened by the velocity of the horizontal motion, and the mercury falls. When the air is fullest of vapours, the mercury falls; the pressure of the atmosphere depending not only on the weight of the fluid, but also on the agility and elasticity of the column of air which is broken and interrupted by such a quantity of moisture floating between, condensing, and ready to fall. These, and many other variations which might be mentioned, are the necessary results of meteors, vapours, and air intermixed in separate portions, and acting with reciprocal, but generally very different powers.

It has frequently been remarked, that the winds in the upper region of the air, as may be seen by the motion of the clouds, are very different from those near the surface of the earth.*

A sign of a change of weather which seemed new and singular to Mr. Borlace,† was thus. August 15, 1752, the wind being at west-north-west, the sky cloudy, the mercury moving upward in the barometer, at about six in the evening, there appeared in the north-east the frustrum of a rain-bow. All the colours were lively and distinct. They call it in Cornwall a weather-dog, or weather's eye, and pronounce it a certain sign of hard rain. The mercury fell

* Memoires de l'Academie Royale des Sciences, pour l'an 1717.

† Natural History of Cornwall, p. 17.

$\frac{2}{10}$, and that without rain. Next morning was dry, but not clear: about eleven it began to rain gently, and at one a flood of rain came on, which continued all night and till the next morning.

Our *northerly* winds in the beginning of the winter may arise from the weight of the cold northern air overcoming the warmer southern air, which, as the heat lessens, is less loaded with vapours, and therefore more easily gives way to the cold northern and denser air. Hence the frequency of north-west winds at that season.

The most general cause of the *easterly* winds in the spring and beginning of summer, arises from the melting of the snow on the continent, as observed by Lord Bacon. The warmth which constantly obtains in a thaw, raises not only much of the melting snow into the air, but the exhalations which had been so long confined by the frost, rise copiously into the air, and become the cause of our easterly winds, which are observed to blow more or less in proportion to the duration and severity of the winter on the continent.

Without entering, however, into the causes of the frequent changes of our winds, concerning which philosophers are by no means agreed, we shall now proceed to the shepherd's rules relating to the winds.

XVII. *Observe that in eight Years Time there is as much South-West Wind, as North-East, and consequently as many wet Years as dry.*

Remarks.

This is, confessedly, a very extraordinary aphorism from a country shepherd, but at the same time it perfectly corresponds with the observations of Dr. Hooke, Dr. Derham, Dr. Grew, and other able naturalists, who with unwearied pains and diligence have calculated the quantity of rain falling in one year, and compared it with that which fell in another. The ingenious observations and calculations of Mr. Kirwan, already noticed,* certainly afford abundant ground to conclude that there is a kind of

* See pages 61—68.

natural balance established, of wet and dry weather, as well as of light and darkness, heat and cold, and similar variations.

It may not be amiss, however, to caution the reader against a mistake into which the manner of this rule being stated may easily lead him: viz. that south-west winds cause rain, and north-east winds fair weather, which is by no means clear or certain. Generally speaking, it is indeed true, that south-west winds and rain, north-east winds and fair weather, come together; but the question is, which causes the other?—and a more difficult question cannot easily be stated, because there seems to be facts on both sides. South-west winds seldom continue long without rain; this seems to prove the affirmative: but on the other hand, when in hard weather rain begins to fall, the wind commonly veers to the south-west; this looks as if the rain caused the wind.

But there is one thing which seems strongly to confirm the shepherd's observation, viz. that in any given place the quantity of rain one year with another is found to be the same by experience; according to which, the following table has been calculated, for the mean quantity of rain falling one year with another in those places which are mentioned, and on this proportion the other seems to be founded.

At	Harlem	24 Inches
	Delf	27
	Dort	40
	Middleburg	33
	Paris	20
	Lyons	37
	Rome	20
	Padua	37½
	Pisa	34¼
	Ulm	27
	Berlin	19½
In	Lancashire	40
	Essex	19½

XVIII. *When the Wind turns to North-East, and it continues two Days without Rain, and does not turn South the third Day, nor Rain the third Day, it is likely to continue North-East for eight or nine Days,* all fair; *and then to come to the South again.*

Remarks.

These rules of our shepherd are among the most valuable of the whole collection; his observation of the manner, which the winds settle in the east or south-west, is particularly worthy the farmer's attention, because it will lead him to most useful fore-knowledge. It is however proper to observe, that as great part of England is a champaign country, at least free from high hills, the winds and weather are more regular there than in mountainous countries, or where the coast is intersected by arms of the sea. The shepherd's remarks, made in the middle of that delightful plain which constitutes the greatest part of England, will therefore not hold so true in other places differently situated.

When he tells us, that in eight years we have as many wet as dry, he does not ascertain what winds bring rain or fair weather; and, as Mr. Worlidge observes, that wind which brings rain to one part of the island, may not to another: for on which coast the sea is nearest, the wind more frequently brings rain to that place, than to another, where the sea is more remote. Therefore it is necessary, that all such as expect any success to their observations, should adjust the rules to the place where they live, and not trust to the observations of other places.

XIX. *After a Northerly Wind for the most part of two Months or more, and then coming South,* there are usually three or four fair Days at first, and then on the fourth or fifth Day comes Rain, *or else the Wind turns North again,* and continues dry.

XX. *If it turn again out of the South to the North-East with Rain, and continues in the North-East two Days without Rain, and neither turns South nor rains the third Day,* it is likely to continue North-East two or three Months.

The wind will finish these turns in three weeks.

XXI. *If it returns to the South within a Day or two without Rain, and turns Northward with Rain, and returns to the South in one or two Days as before, two or three times together after this sort,* then it is like to be in the South or South-West two or three Months together, *as it was in the North before.*

The winds will finish these turns in a fortnight.

XXII. *Fair Weather for a Week with a Southern Wind,* is like to produce a great Drought, *if there has been much Rain out of the South before. The Wind usually turns from the North to South with a quiet Wind without Rain; but returns to the North with a strong Wind and Rain. The strongest Winds are when it turns from South to North by West. When the North Wind first clears the air, which is usually once a Week,* be sure of a fair Day or two.

XXIII. *If you see a Cloud rise against the Wind, or Side-wind, when that Cloud comes up to you,* the Wind will blow the same way the Cloud came. *The same rule holds of a clear place, when all the Sky is equally thick, except one clear Edge.*

Southerly and westerly winds generally prove rainy in this island, there being so great an extent of sea to the south-west: yet places far distant from that sea, or which are screened from it by high mountains, have fair weather; as is the case on the north-east coast of Scotland, where the vapours are intercepted by the Grampian hills. The easterly winds, coming to the south part of the island over a narrow tract of sea, are generally fair, except in winter, when they bring on the dark, heavy sky. They are extremely sharp and cold in the winter, coming from a frozen continent; but if inclined to the south, are hot and dry in the summer, as coming from the continent then heated by the sun. The easterly winds crossing a much wider sea in their passage to Scotland, prove generally rainy all along the east of that country; but fair on the west. We may easily conceive that the air, in crossing the German ocean, may take up water enough to cause this rain by its faculty of attracting water, before mentioned.

A wind blowing from the sea is observed to be always moist; cold in summer, and warm in winter, unless the sea be frozen up: (*i. e.* the temperature of wind blowing over water is more equal than that of wind blowing over land:) and winds blowing from large continents are dry, warm in summer, and cold in winter. If the frost is so great as to freeze the vapour as it rises from the sea, it must feel extremely sharp and cold to our bodies; though by the thermometer the cold may be the same as in lofty situations, to which such heavy vapours seldom ascend in winter.

Lord Bacon observes, that "when the wind changes comformable to the motion of the sun, that is from east to south, from south to west, &c. it seldom goes back; or if it does, it is only for a short time: but if it moves in a contrary direction, *viz.* from east to north, from north to west, it generallyr eturns to the former point, at least before it has gone quite through the circle. When winds continue to vary for a few hours, as if it were to try in what point they should settle, and afterwards begin to blow constant, they continue for many days. If the south wind begins to blow for two or three days, the north wind will blow suddenly after it: but if the north wind blows for the same number of days, the south will not rise till after the east has blown a while. Whatever wind begins to blow in the morning, usually continues longer than that which rises in the evening."

Mr. Worlidge observes, that "if the wind be east or north-east in the fore part of the summer, the weather is likely to continue dry: and if westward towards the end of the summer, then will it also continue dry. If in great rains the winds rise or fall, it signifies that the rain will forthwith cease. If the colours of the rain-bow tend more to red than any other colour, wind follows; if green or blue predominate, then rain."

SECTION VI.

PROGNOSTICS OF THE WEATHER, TAKEN FROM THE SEASONS.

XXIII. SPRING and SUMMER. *If the last eighteen Days of February and ten Days of March be for the most part rainy, then the Spring and Summer Quarters are like to be so too: and I never knew a great Drought but it entered in that Sesaon.*

Remarks.

Observation will easily discover whether this rule be well or ill founded, that is to say, whether our shepherd's observation will serve for other places or not, and where it will serve and where not. It is but highly probable, that the weather in one season of the year determines the weather in another: for instance, if there be a rainy winter, then the autumn will be dry; if a dry spring then a rainy winter. Our forefathers had abundance of odd sayings upon this subject, and some proverbs for every month in the year, many of which seem to have but indifferent foundations. There can however be no harm in observing them, in order to discover whether these traditional remarks are well or ill founded.

Janiver freeze the pot by the fire.
If the grass grow in Janiver,
It grows the worse for't all the year.
The Welchman had rather see his dam on the bier

Than to see a fair Februeer.
March wind and May sun
Make clothes white and maids dun.
When April blows his horn,
It's good both for hay and corn.
An April flood
Carries away the frog and her brood.
A cold May and a windy
Makes a full barn and a findy.
A May flood never did good.
A swarm of bees in May
Is worth a load of hay.
But a swarm in July
Is not worth a fly, &c.

XXIV. Winter. *If the latter End of October and Beginning of November be for the most part warm and rainy, then January and February are like to be frosty and cold, except after a very dry Summer.*

XXV. *If October and November be Snow and Frost, January and February are likely to be open and mild.*

Remarks.

The reason of this observation, supposing it to be true, is to be sought in that balance of the weather which Providence has established. There is not only a time to sow, and a time to reap, but there is a time also for dry and a time for wet weather; and if these do not happen at proper seasons, they will certainly happen at others: for not only has the wisdom of philosophers discovered, but their experiments and observations have demonstrated, that there is a certain rule or proportion observed between wet weather and dry in every country, so that it is nearly the same in every annual revolution; neither is it wet and dry weather only, but hot and cold, open and frost, that are thus regulated. Hence we see, that when the Scripture represents to us God's settling things by weight and measure, it speaks not only elegantly, but exactly: for we are not to understand, by Providence, any extraordinary or supernatural interposition of almighty power, but the constant and settled order established by the will of that Almighty Beiug, which order we ordinarily call Nature.

The following rules, laid down by Lord Bacon, will conclude our remarks on the shepherd's prognostications of the changes of the weather from the seasons.

If the wainscot or walls that used to sweat be drier than usual, in the beginning of winter, or the eves of houses drop more slowly than ordinary, it portends a hard and frosty winter: for it shews an inclination in the air to dry weather, which, in winter, is always joined with frost.

Generally, a moist and cool summer portends a hard winter.

A hot and dry summer and autumn, especially if the heat and drought extend far into September, portend an open beginning of winter, and cold to succeed towards the latter part of the winter, and beginning of spring.

A warm and open winter portends a hot and dry summer; for the vapours disperse into the winter showers; whereas cold and frost keep them in, and convey them to the late spring and following summer.

Birds that change countries at certain seasons, if they come early, shew the temper of the weather, according to the country whence they came; as, in the winter, wood-cocks, field-fares, snipes, &c. if they come early, shew a cold winter; and the cuckoos, if they come early, show a hot summer to follow.

A serene autumn denotes a windy winter; a windy winter, a rainy spring; a rainy spring, a serene summer; a serene summer, a windy autumn; so that the air, on a balance, is seldom debtor to itself; nor do the seasons succeed each other in the same tenor for two years together.

In addition to these rules, Mr. Worlidge remarks, that

If at the beginning of the winter the south wind blow, and then the north, it will probably be a cold winter; but if the north wind first blow, and then the south, it will be a warm and mild winter.

If the oak bear much mast, it prognosticates a long and hard winter. The same has been observed of hips and haws. If broom be full of flowers, it usually signifieth plenty.

Mark well the flowering almonds in the wood;
If od'rous blooms the bearing branches load,
The glebe will answer to the sylvan reign,
Great heats will follow, and large crops of grain.
But if a wood of leaves o'ershade the tree,
Such and so barren will the harvest be.
In vain the hind shall vex the threshing floor,
For empty chaff and straw will be thy store.

VIRGIL.

This observation, says Mr. Worlidge, hath proved for the most part true for several years now past; as in 1673 and 1674 there were but few nuts, and cold and wet harvests; in 1675 and 1676, were plenty of nuts, and heavy and dry harvests; but more especially in 1676 was a great shew of nuts, and a very hot and dry harvest succeeded.

APPENDIX.

MISCELLANEOUS OBSERVATIONS,

NOT REFERIBLE TO ANY OF THE PRECEDING PARTS OR SECTIONS.

I. *Observations on the Winds.*

In the former part of this work, we have stated the various states of weather indicated by the blowing of particular winds; in addition to those remarks, we annex the following particulars, which will be found not devoid of interest to the attentive observer of nature.

Wind, it may be observed, is a sensible agitation of the atmosphere, caused by a quantity of air blowing from one place to another. As not only navigation depends in a great degree upon the direction and force of the winds, but also the temperature of climates and the healthiness of the atmosphere are materially influenced by them, the following facts, drawn from attentive observation, are submitted to the reader's attention.

---

1. *Sea and Land Breezes.*

Sea-breezes commonly rise in the morning about nine o'clock. They first approach the shore gently, as if they were afraid to come near it. The breeze comes in a fine, small, black curl upon the water, whereas all the sea between it and the shore (not yet reached by it) is as smooth and even as glass in comparison. In half an hour's time after it has reached the shore, it fans pretty briskly, and increases gradually till twelve o'clock; then it is commonly the strongest, and lasts so till two or three, a very brisk gale.—After three, it begins to die away again, and gradually withdraws its force till all is spent; and about five o'clock it is lulled asleep, and comes no more till next morning.

As the sea-breezes blow in the day, and rest in the night; so, on the contrary, the land-breezes blow in the night, and rest in the day, alternately succeeding each other: they spring up between six and twelve at night, and last till six, eight, or ten in the morning.

---

2. *The Trade-Winds.*

The trade-winds denote certain regular winds at sea, blowing either constantly the same way, or alternately this way and that: they are thus designated from their use in navigation and in the Indian trade.

The constant trade-winds do not usually blow near the shore, but only on the ocean, at least 30 or 40 leagues off at sea, clear from any land, especially on the west coast, or side of any continent: for, on the east side, the easterly wind being the true trade-wind, blows almost home to the shore, so near as to receive a check from the land-wind.

And not only the general trade-winds, but also the constant coasting trade-winds, are in like manner affected by the lands, as is proved on the coast of Angola and Peru. But it must be remarked, that the trade-winds which blow on any coast, except the north coast of Africa, whether they are constant and blow all the year, or whether they are shifting winds, do never blow right *in* on the shore, nor right *along* shore, but go *slanting*, making an acute angle of about 22 degrees. Therefore, as the land tends more east or west, from north or south on the coast; so the winds do alter accordingly.

---

3. *Direction of the Winds.*

From an average of ten years of the register kept by the order of the Royal Society, it appears, that at London the winds blow in the following order:

Winds.	*Days.*
South-west - - -	112
North-east - - -	58
North-west - - -	50
West - - - -	53
South-east - - - -	32
East - - - - -	26
South - - - -	18
North - - - -	16

It appears from the same register, that the south-west wind blows at an average more frequently than any other wind during every month of the year, and that it blows longest in July and August; that the north-east blows most constantly during January, March, April, May, and June, and most seldom during February, July, September, and December; and that the north-west wind blows oftener from November to March, and more seldom during September and October, than any other months. The south-west winds are also most frequent at Bristol, and next to them are the north-east.*

* *Philosoph. Trans. of the Royal Society*, vol. lvi. p. 683.

In Ireland the south-west and west are the grand trade-winds, blowing most in summer, autumn, and winter, and least in spring. The north-east blows most in spring, and nearly double to what it does in autumn and winter. The south-west and north-west are nearly equal, and are most frequent after the south-west and west.*

The direction in which the wind blows may be ascertained by observing certain flowers, some kinds of which are adapted both to the winds and to rain. Of this description are the flowers of *peas*, which are furnished with small boats to cover and shelter the stamina, and the embryos of their fruits. Further, they have large pavilions, and rest on tails bent, and elastic like a nerve; so that when the wind blows over a field of peas, all the flowers may be seen to turn their backs to the wind, like so many weather-cocks.

* Dr. Rutty's History of the Weather in Dublin, &c.

II. *Observations on Lightning.*

Lightning strikes the highest and most pointed objects in its course, in preference to others, as hills, trees, spires, masts of ships, &c. So, all pointed conductors receive and throw off the electric fluid more readily than such as are terminated by flat surfaces. Lightning is observed to take and follow the readiest and best conductor.

With regard to places of safety in times of thunder and lightning, Dr. Franklin's advice is, to sit in the middle of a room, provided it be not under a metal lustre suspended by a chain, sitting on one chair, and laying the feet on another. It is still better, he says, to bring two or three mattresses, or beds, into the middle of the room, and folding them double, to place the chairs upon them; for, as they are not so good conductors as the walls, the lightning will not be so likely to pass through them. But the safest place of all is in a hammock hung by silken cords, at an equal distance from all the sides of the room. Dr.

Priestley observes that the place of most perfect safety must be the cellar, and especially the middle of it; for when a person is lower than the surface of the earth, the lightning must strike it before it can possibly reach him. In the fields, the place of safety is within a few yards of a tree, but not quite near it.

III. *Water-spouts.*

A water-spout is an extraordinary meteor, most frequently observed at sea. It commonly begins by a cloud which appears very small, and which sailors term the *squall:* this in a little time augments into an enormous cloud of a cylindrical form, or that of a reversed cone, and produces a noise somewhat like an agitated sea, sometimes accompanied with thunder and lightning, and also pouring down large quantities of hail or rain, sufficient to inundate large vessels, to overwhelm trees and houses, and every thing which opposes its violent impetuosity.*

Water-spouts are more frequent at sea than by land: and so convinced are mariners of their dangerous consequences, that when they perceive their approach, they frequently endeavour to dissipate them by firing a cannon, before they approach too near the ship. Water-spouts have also been known to have committed great devastations by land; although where there is no water near, they generally assume the form of a whirlwind.

Various extraordinary effects have been recorded, as being produced by water-spouts, the descriptions of which most probably have been much exaggerated. One at Topsham, in Devonshire, is said to have cut down an apple tree, several inches in diameter: another, it is said, seemed to be produced by a concourse of winds, turning like a screw, the clouds dropping down into it; it threw down trees and branches, with a gyratory or circular motion.

* Gregory's Economy of Nature, vol. 1. p. 370

IV. *Nautical Observations may be made from Aquatic Plants.*

The seeds of aquatic plants have forms no less adapted than those of their leaves, to the places where they are destined to grow; they are all constructed in a manner most proper for sailing. Some of them are fashioned like shells; others like boats, rafts, and skiffs, as well as single and double canoes, similar to those of the South Seas. By an attentive study of this part alone of natural history, a great number of very curious discoveries might be made respecting the art of crossing currents of every sort. A very ingenious writer concludes this observation by the following remark. "I am persuaded that the first men, who were much better observers than we are, took their different methods of travelling by water from those models of nature, of which, with all our pretensions to discovery, we are but feeble imitators."

The aquatic or maritime pine has its kernels inclosed in a kind of small bony shoes, notched on the lower side, and covered on the upper with a piece resembling a ship's hatch. The walnut, which delights so much in the banks of rivers, has its fruit contained in two small boats, fitted to each other. The hazel, which becomes so bushy on the brink of rivulets, and the olive, which loves the sea-shore to such a degree that it degenerates in proportion as it is removed from it, bear their seed inclosed in a species of small casks, capable of enduring the longest voyages. The red berry of the yew, whose favourite residence is the cold and humid mountains, near the margin of lakes, is hollowed out into a little bell. This berry, on dropping from the tree, is at first carried down by its fall to the bottom of the water; but it instantly returns to the surface by means of a hole, which nature has contrived in the form of a navel, above the seed. In this aperture is lodged a bubble of air, which brings it back to the surface of the water, by a mechanism more ingenious than that of the diving bell, as the vacuum of the latter is undermost, and that in the berry of the yew uppermost.

V. *Miscellaneous Observations on Plants.*

In the former part of this work, notice has been taken of the indications of weather afforded by the vegetable creation: beside affording these prognostics, many plants also fold themselves up at particular hours, and with such regularity, as to have acquired particular names from this property. The following are among the more remarkable plants of this description.

1. The Goat's Beard, or Tragopogon of Linnæus: the flowers of this plant open in the morning at the approach of the sun, and (without regard to the state of the weather) regularly shut about noon. Hence it is generally known in the country by the name of *John-go-to-bed-at-noon.*

2. The Princesses' Leaf, or Four o'Clock Flower, in the Malay Islands, is an elegant shrub so called by the natives, because their ladies are fond of the grateful odour of its white leaves. It takes its generic name from its quality of opening its flowers at four in the evening, and

not closing them in the morning till the same hour returns, when they again expand in the evening at the same hour.

Many people transplant them from the woods into their gardens, and use them as a *dial* or *clock*, especially in cloudy weather.

3. THE EVENING PRIMROSE. This flower is well known from its remarkable properties of regularly shutting with a loud popping noise, about sun-set in the evening, and opening at sun-rise in the morning. A curious observer may receive pleasure by noticing how regularly, after six o'clock, these flowers will report the approach of night.

4. The *Parkinsonia* or TAMARIND TREE, the *Lapsana* or NIPPLE-WORT, *Nymphæa* or WATER LILY, *Calendula* or MARIGOLD, *Æschynomene* or BASTARD SENSITIVE PLANT, and several others of the *Diadelphia* class, in serene weather expand their leaves in the day-time, and contract them during the night. According to some botanists, the Tamarind-tree enfolds within its leaves the flowers or fruit, every night, in order to guard them from cold or rain.

5. The flower of the GARDEN LETTUCE, which is in a vertical plane, opens at seven o'clock, and shuts at ten.

6. A species of serpentine ALOES, without prickles, whose large and beautiful flower exhales a strong odour of the vanilla, during the time of its expansion, which is very short, is cultivated in the imperial garden at Paris. It does not blow till towards the month of July, and about five o'clock in the evening; at which time it gradually opens its petals, expands them, droops, and dies. By ten o'clock the same night, it is totally withered, to the great astonishment of the spectators, who flock in crowds to see it.

In like manner, the attentive observer of nature may notice, how almost every species of flowers are expanded or opened by the genial rays of the sun; but in the evening and during a moist state of the air the flowers close, or contract, lest the moisture (penetrating the dust of the anthers) should coagulate the same, and prevent it from being blown on the stigmata or summits. A very remarkable circumstance, attending plants of this class is, that when the fecundation is completed, the flowers do not contract either in the day, or in the evening, nor at the approach of rain.

In addition to the flowers of plants above mentioned, as closing and opening their petals at certain hours of the day, many others might be specified. The illustrious Linnæus has enumerated forty-six flowers, which possess this kind of sensibility: he divides them into three classes.

1. *Meteoric Flowers*, which less accurately observe the hour of folding, but are expanded sooner or later according to the cloudiness, moisture or pressure of the atmosphere.

2. *Tropical Flowers*, that open in the morning and close before evening every day; but the hour of their expanding becomes earlier or later as the length of the day increases or decreases.

3. *Equinoctial Flowers*, which open at a certain and exact hour of the day, and for the most part close at another determinate hour. *

Hence the Horologe, or BOTANICAL WATCH, is formed from numerous plants, which are of frequent occurrence in this country. (See the FRONTISPIECE.)

* Dr. Darwin's "Poetical Works," vol. II. p. 90, *note*.

In the engraving just referred to, the letters M.N. E. and A. in the *third* and *sixth* circular spaces of the Dial, signify Morning, Night, Evening, and Afternoon; and the figures on the *fourth* and *seventh* spaces refer to the hour of the day or night, &c. the flowers open or shut at. For example, let the reader look for *Papaver nudicaule*, or naked poppy, and he will find in the corresponding spaces, that it opens in the morning at five o'clock, and closes in the evening at seven. The other letters and figures will correspond in a similar manner with the flowers mentioned in the *first* or *large* space of the dial. These observations, however, can only be made during the summer months, but they will afford ample and rational amusement to those who are partial to botanical experiments.

Three of the Cardinal Points of the Compass, shown by the Sun-flower.

The common annual sun-flower has the remarkable property of following the course of the sun during the day; and in the night (the stalks untwisting) returns to the east, to face the sun next morning. In July, the top of the sun-flower being tender, and the flower near beginning to blow—if the sun rise clear, the flower faces towards the east, and the sun continuing to shine, at noon it faces to the south, and at six in the evening to the west. This is not by turning round with the sun, but by nutation; the cause of which is, that the side of the stem next the sun perspiring most, it shrinks, and this plant perspires much.

VI. *Important Remark on the Benefit of* EARLY RISING, *by the late Rev. Dr.* DODDRIDGE.

"The difference between rising at five, and at seven o'clock in the morning, for the space of forty years, supposing a man to go to bed at the same hour at night, is *nearly equivalent to the addition* of TEN YEARS *to a man's life*."*

* *Family Expositor*, on Rom. XIII. 13, *note* k.

THE END.

INDEX.

Harding and Wright, Printers, St. John's Square, London.

Printed in the United States
By Bookmasters